Mass Spectrometry for the Novice

Mass Spectrometry for the Novice

John Greaves and John Roboz

CRC Press
Taylor & Francis Group
Boca Raton London New York

CRC Press is an imprint of the
Taylor & Francis Group, an **informa** business

CRC Press
Taylor & Francis Group
6000 Broken Sound Parkway NW, Suite 300
Boca Raton, FL 33487-2742

© 2014 by Taylor & Francis Group, LLC
CRC Press is an imprint of Taylor & Francis Group, an Informa business

No claim to original U.S. Government works

Printed on acid-free paper
Version Date: 20130514

International Standard Book Number-13: 978-1-4200-9418-3 (Paperback)

Library of Congress Cataloging-in-Publication Data

Greaves, John (Director of Mass Spectrometry Facility)
  Mass spectrometry for the novice / John Greaves and John Roboz.
    pages cm
  Includes bibliographical references and index.
  ISBN 978-1-4200-9418-3 (pbk.)
  1. Mass spectrometry. I. Roboz, John, 1931- II. Title.

QD96.M3G74 2013
543'.65--dc23
                                                                    2013009379

Visit the Taylor & Francis Web site at
http://www.taylorandfrancis.com

and the CRC Press Web site at
http://www.crcpress.com

Dedication

To my family, especially my wife, Izzy, and my father.

J.G.

To my granddaughters, Alexandra and Melina,
with love and respect.

J.R.

Contents

Supplementary Resources Disclaimer

Additional resources were previously made available for this title on CD. However, as CD has become a less accessible format, all resources have been moved to a more convenient online download option.

You can find these resources available here: http://resourcecentre.routledge.com/books/9781420094183

Please note: Where this title mentions the associated disc, please use the downloadable resources instead.

List of Figures

List of Tables

Acknowledgments

Our thanks are due to several individuals for reading various parts of the text and making thoughtful and relevant comments and suggestions: Dr. Ben Haffensteiner, Dr. Martin Schnermann, James Jernee, Dr. Beniam Berhane, Dr. Sool Cho, and Tony Hickson. In particular, we acknowledge Dr. Barrie Peake from the University of Otago in Dunedin, New Zealand. Barrie was extremely willing and diligent in looking over the text both scientifically and with a language eye as well as providing us with comments and suggestions that have led to many improvements. We wish to thank Ms. Fang Q, a genuine novice in mass spectrometry, for reading the manuscript and offering her thoughts.

We would like to express very special thanks to our respective wives, Izzy and Julia, for their understanding, encouragement, and patience—in summary, for just tolerating us (barely).

About the Authors

John Greaves is director of the Mass Spectrometry Facility in the Department of Chemistry, University of California, Irvine. He received his BSc in zoology and genetics from Leeds University in 1975 and a PhD from Liverpool University/Liverpool School of Tropical Medicine in 1979 for research on the metabolism of the antimalarial drug primaquine.

On moving to the United States he spent 5 years at Mount Sinai School of Medicine and 7 years at the Virginia Institute of Marine Science, where he was involved with cancer and environmental research, respectively. He moved to the University of California, Irvine in 1992 where he has been particularly interested in the use of open access mass spectrometry to enable scientists lacking experience in the technique to facilitate their research by obtaining mass spectrometric data rapidly on a 24/7 basis. He has over 80 publications.

John Roboz is a professor in the Department of Medicine (Division of Hematology/Medical Oncology) of the Icahn School of Medicine at Mount Sinai, New York. He holds a BS degree (1955) from Eotvos University, Budapest, Hungary, and MS (1960) and PhD (1962) degrees in physical chemistry from New York University.

After immigrating to the United States from Hungary in 1957, he worked as a senior research chemist in industrial research laboratories (General Telephone and Electronics, and Air Reduction Company). In 1969, he joined the Mount Sinai School of Medicine as an associate professor and in 1980 he became a professor.

His primary research interest has been the development of mass spectrometric techniques in analytical pharmacology, and biochemistry of new antineoplastic agents, and their applications in collaborative studies with basic scientists and clinicians.

Dr. Roboz has more than 135 publications. His first book, *Introduction to Mass Spectrometry: Instrumentation and Techniques* (Wiley, 1968), was reprinted in 2000 by the American Society for Mass Spectrometry in the *Classic Works in Mass Spectrometry* series. His second book, in 2002, *Mass Spectrometry in Cancer Research* (CRC Press), won a commendation award from the British Medical Association.

The authors have been long-term collaborators since the 1980s when they started to work on the occurrence of polybrominated biphenyls in the population of Michigan resulting from an industrial accident. J.G. spent extended amounts of time asking J.R. (to his mind) innumerable questions while correcting and proofreading his second book. The current effort began with the suggestion from J.G. that a somewhat different approach should be taken to the introduction of mass spectrometry to an expanding audience of novices with no training in the technique, and to those who want to expand their knowledge.

Abbreviations

APCI	Atmospheric pressure chemical ionization
API	Atmospheric pressure ionization
APPI	Atmospheric pressure photo-ionization
B	Magnetic analyzer
CAD	Collisionally activated dissociation/decomposition
CE	Capillary electrophoresis
CI	Chemical ionization
CID	Collisionally induced dissociation
Da	Dalton (unit of atomic mass)
DART	Direct analysis in real time
DDA	Data dependent acquisition
DESI	Desorption electrospray ionization
DIA	Data independent acquisition
ECD	Electron capture dissociation
ECNI	Electron capture negative ionization
EI	Electron ionization
EM	Electron multiplier
ESI	Electrospray ionization
ESA	Electrostatic analyzer
ETD	Electron transfer dissociation
eV	Electron volt
FAB	Fast atom bombardment
FT	Fourier transform
FT-ICRMS	Fourier transform ion cyclotron resonance mass spectrometer
FT-MS	Fourier transform mass spectrometer
FWHM	Full width (at) half maximum (height) (resolution definition)
GC	Gas chromatograph, gas chromatography
GC-MS	Gas chromatography–mass spectrometry
HPLC	High-pressure/performance liquid chromatography
ICR	Ion cyclotron resonance
IMS	Imaging mass spectrometry
IMS	Ion mobility separator
IRMPD	Infrared multiphoton dissociation
IT	Ion trap
kDa	Kilodalton
LC	Liquid chromatograph, liquid chromatography
LC-MS	Liquid chromatography–mass spectrometry
LIT	Linear on trap
m	Mass
MALDI	Matrix-assisted laser desorption/ionization

MCP	Microchannel plate (detector)
mDa	Millidalton
mmu	Millimass unit
MRM	Multiple reaction monitoring (rarely used)
MS	Mass spectrometer, mass spectrometry, mass spectrum
MS/MS	Mass spectrometry/mass spectrometry, tandem mass spectrometry
MS^n	Multistage mass spectrometry/mass spectrometry (n usually 3)
MW	Molecular weight (used interchangeably with molecular mass)
m/z	Mass/charge
NCI	Negative chemical ionization
PCI	Positive chemical ionization
PI	Photo-ionization
ppb	Parts per billion
ppm	Parts per million
PSD	Post-source decay
q	Quadrupole, rf only (for transmission and focusing only)
Q	Quadrupole (analytical)
QIT	Quadrupole ion trap
QqQ	Triple quadrupole
QTOF	Quadrupole time-of-flight
rf	Radio frequency
SELDI	Surface-enhanced laser desorption ionization
SIM	Selected ion monitoring
SRM	Selected reaction monitoring
TIC	Total ion chromatogram
TIC	Total ion current
TOF	Time-of-flight
TOF/TOF	Time-of-flight/time-of-flight
UV	Ultraviolet
z	Charge of an ion (in integer multiples, electron or proton)

d	Deca, 10
c	Centi, 10^{-2}
m	Milli, 10^{-3}
μ	Micro, 10^{-6}
n	Nano, 10^{-9}
p	Pico, 10^{-12}
f	Femto 10^{-15}
a	Atto 10^{-18}
z	Zepto 10^{-21}

REFERENCE

IUPAC. Standard definitions relating to mass spectrometry. http://mass-spec.lsu.edu/msterms/index.php/Main_Page. Started with provisional definitions in 2006. There are 795 entries in "Mass Spectrometry Terms" as of February 2013.

What a way to start!
Is it true that even the title of
this book is a misnomer?
Is it true that we do not
measure mass in a mass
spectrometer?
Is it true that we measure
m/z - whatever that is -
(or a related property)?

To the Reader

There are two definitions for the word *novice*. The first is that a novice is a person new to or inexperienced in a field. Yes, this book is addressed to novices in mass spectrometry. A novice is also described as a person who has entered a religious order and is under probation before taking vows. Perhaps this is a stretch, but many mass spectrometrists (particularly the older ones) believe that the field (or art?) of mass spectrometry is not unlike a religious order—once entered, you are in it for keeps. And when is your probation over? What do you have to know before you are confirmed? When is a novice no longer a novice? Are we all novices forever because there is always something more to comprehend, to learn, and then apply? Do even the high priests of the discipline understand the mechanism of electrospray ionization? Such meditations are probably way too philosophical. So let's say to the novice mass spectrometrists: please just learn what's in this book, and you will be ready to take your vows.

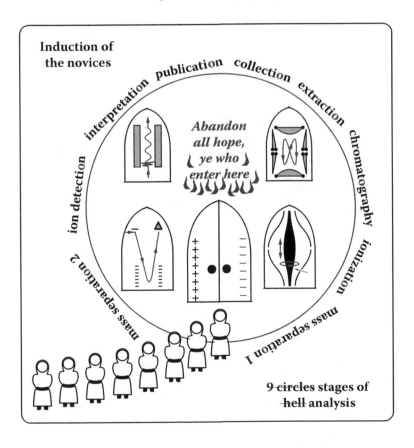

Mass spectrometry has grown immeasurably over the last 20 years. It may well be the technique to turn to if you want to analyze gases, pharmaceuticals, drugs of abuse, or environmental pollutants, search for explosives at airports, sequence peptides from proteins, or if you wish to understand the role of complex carbohydrates in cancer. There are an ever-increasing number of scientists, physicians, and technicians who are coming into contact with the technique. Most individuals do not need an in-depth understanding of the complex physics and chemistry behind mass spectrometry but would benefit significantly from an understanding of the basic concepts and the processes that occur in these instruments.

We have attempted to provide straightforward explanations and readily understandable figures and schematics (all have been drawn specifically for this book). Efforts have been made to keep the language simple, and while this may not be a late-night page turner, we hope that reading this book will be more like listening to a favorite lecturer. The book is not intended to be an exhaustive treatise (there are several available). We have, instead, concentrated on explaining the operation of the components of the instruments, alone and in various combinations, and the commonly used techniques and strategies. It is our intention to help novices make informed choices that are appropriate to the objectives of their tasks and (hopefully) the logic behind assigned projects.

Mass spectrometers are versatile instruments that can be used to analyze, both qualitatively and quantitatively, an amazing range of compounds from simple gases to complex biopolymers. The multiple ways of combining instrument components into systems enable sophisticated approaches to introduce and analyze samples. There are also extensive software packages for handling the enormous quantity of data generated. The strength of mass spectrometry lies in its extraordinary versatility, but the price of that diversity is that mass spectrometers are often daunting instruments, especially for the new user.

This book is also intended as a resource for educators: the illustrations have been prepared anew, also with an eye to their being projected during lectures or included in handouts. Accordingly, the illustrations are also provided on an enclosed CD.

Each figure is supplied in two formats, one with extensive legends (probably better viewed on a computer screen or on handouts) and the other with shortened legends (for projection).

The book concentrates on the most common instruments used in chemical, environmental, biological, and medical research. Less frequently encountered forms of mass spectrometry are not addressed, such as inductively coupled plasma mass spectrometry (ICP-MS) isotope ratio mass spectrometry (IRMS), accelerator mass spectrometry (AMS), proton transfer reaction mass spectrometry (PTRMS), and single-particle laser ablation time-of-flight mass spectrometry (SPLAT).

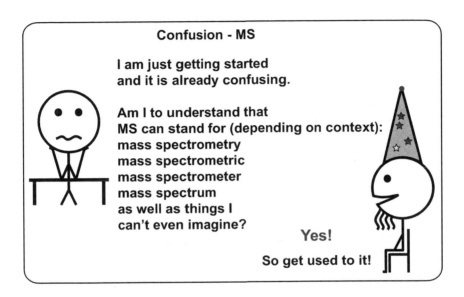

The book is divided into six chapters. Chapter 1 is an overview, including essential definitions, and is intended to assist the novice in pondering what instruments and methods are appropriate for the data they need to obtain. This chapter is intentionally brief and the reader will need to explore later chapters for more detailed descriptions of specific topics. Chapter 2 provides information on the components of instruments and the processes occurring within each part of the various instrument systems. Chapter 3 discusses techniques and strategies and the types of data generated. Chapter 4 includes summaries of a small set of representative applications from published articles aiming to illustrate the impressive diversity of mass spectrometry. Some of these articles are straightforward while others are much more complex, requiring additional effort to comprehend. The reader is encouraged to read the original articles. Chapter 5 contains "the absolute essentials," a set of "take home" messages. Chapter 6 contains resources: lists of recommended books, including a few truly classical books and papers, a set of selected review articles for all major areas discussed, a list of relevant journals, and a number of useful Internet addresses, including tutorials, the perusal of which is highly recommended.

Finally, a comment about the inclusion of cartoons. Although they are not normally used in books like this, we hope that most readers will find them funny (or at least moderately amusing) and, on occasion, educational. Perhaps, as one moves on from novice status, some of them will (magically) become more humorous and relevant.

Now everybody,
memorize and repeat
twice daily this MANTRA
for mass spectrometrists

A mass spectrometer can only analyze IONS
(electrically charged molecules) - and they
must be in the GASEOUS phase

Mass spectrometers do not measure mass
but do MEASURE m/z (mass-to-charge ratios);
z usually, but by no means always, equals 1

There is NO such thing as a "CORRECT" MASS
SPECTRUM for a compound; this is what makes
mass spectrometry so DIVERSE and WONDERFUL

The mass spectrometer NEVER LIES; if you do
not understand the result - do more thinking

1 An Overview

1.1 A BRIEF HISTORY

The origins of mass spectrometry go back to the characterization of "positive rays" by J. J. Thompson in the early 1900s. The technique was in the domain of physicists and physical chemists for the next 60 years, during which there were significant advances in both instrumentation and applications. Among the former were the electron ionization source as well as double-focusing magnetic and time-of-flight analyzers. Applications included the demonstration, by F. W. Aston, that many elements had different isotopic forms. The ability to differentiate isotopes led to the collection of the uranium-235 isotope and its use in the atom bomb.

Most areas of scientific endeavor are characterized by periods of ongoing progressive improvements interrupted by revolutionary changes. Mass spectrometry is no different. The commercialization of the interfacing of gas chromatography with mass spectrometry in the late 1960s was such a revolution. The coupling of the two techniques enabled the separation of complex mixtures so that individual components could be admitted sequentially into the mass spectrometer. Because of the need for samples to be in the vapor phase (much will be said about this later), there remained the problem of what to do with polar compounds that decompose when heated, prior to vaporization. This class of compounds includes the vast majority of molecules of biological interest. Chemical derivatization enabled the analysis of some polar compounds, and the polarity barrier was also partially breached by the development of fast atom bombardment (FAB) in the late 1970s. FAB was discarded in the 1990s when mass spectrometry was revolutionized again, this time by the commercialization of electrospray ionization (ESI) and matrix-assisted laser desorption/ionization (MALDI). ESI facilitated the coupling of liquid chromatography (LC) with mass spectrometry, allowing the on-line analysis of polar, bioactive compounds such as drugs and their metabolites. In addition, and importantly, ESI and MALDI enabled mass spectrometry to move into the field of protein (and other biopolymer) analysis.

Mass spectrometry has earned a number of Nobel Prizes, starting with Joseph J. Thompson, who received the physics prize in 1906 for the development of positive rays. Next, the chemistry award went to Francis W. Aston, in 1922, for his research elucidating the existence of isotopes. Moving to the modern era, Wolfgang Paul and Hans G. Demelt were cited in 1989 for developing ways to trap ions (physics), while John B. Fenn and Koichi Tanaka received the chemistry prize, in 2002, for the development and application of ESI and MALDI, respectively, to the analysis of proteins.

We are inheritors of a storied technique that continues to develop and is used to explore many aspects of our everyday world. While trips to Stockholm by mass spectrometrists are few and far between, each of us who uses these instruments contributes, in some small part, to the story of mass spectrometry.

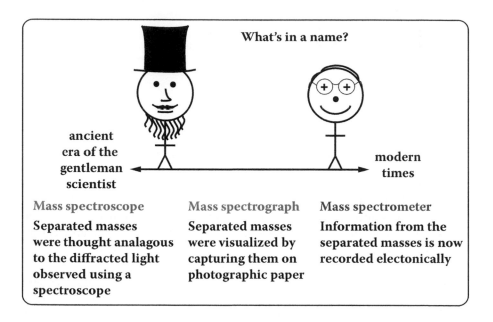

What's in a name?

ancient era of the gentleman scientist

modern times

Mass spectroscope

Separated masses were thought analagous to the diffracted light observed using a spectroscope

Mass spectrograph

Separated masses were visualized by capturing them on photographic paper

Mass spectrometer

Information from the separated masses is now recorded electonically

1.2 DEFINITIONS CONCERNING INSTRUMENTS, MASS, *m/z*, AND IONS

These definitions are intended to introduce and describe the essential concepts of mass spectrometry in as elementary a manner as is reasonable. The definitions given are practical and are not meant to be rigorous. Apologies are made in advance for definitions that are not completely clear or appear confusing. Regretfully, just like in many other areas of science, confusion often reigns concerning even the most elementary (and essential) notions and designations. All subsequent sections include extended descriptions of instrumentation, experiments, relevant examples, and contexts for these definitions.

What is a mass spectrometer? A *mass spectrometer* is an analytical instrument that produces a beam of gas phase ions from samples (*analytes*), sorts the resulting mixture of ions according to their *mass-to-charge* (*m/z*) ratios using electrical or magnetic fields (or combinations thereof), and provides analog or digital output signals (*peaks*) from which the mass-to-charge ratio and the *intensity* (abundance) of each detected ionic species may be determined.

All mass spectrometers, whatever the level of their sophistication, operate on the premise that the sample is ionized, and the resulting ions are separated according to a parameter related to their mass-to-charge ratio and then detected (Figure 1.1).

What are the major components of a mass spectrometer? The major components that make up a mass spectrometer serve to carry out the processes of ionization, mass separation, and detection shown in Figure 1.2. There are six major components: (1) sample introduction system, (2) ion source where the analytes are vaporized and ions are produced, (3) mass analyzer where the ions are separated according to their

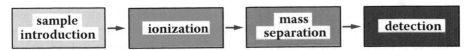

Basic sequence of events in a mass spectrometer

FIGURE 1.1   Basic sequence of events in a mass spectrometer.

m/z ratios, (4) ion detector where the signal intensities of each separated *m/z* value are determined, (5) vacuum system, needed to prevent the loss of ions through collisions with neutral gas molecules as well as with the walls of the mass analyzer, the detector, and sometimes the ion source, and (6) computers (one or more), to control the operation of the instrument, record, and process the data generated (Figure 1.2).

Each component comes in a variety of forms and, on occasion, some components can be combined, e.g., sample introduction and ionization. All components will be outlined in Chapter 1 and discussed throughout the text both in terms of how they operate (Chapter 2) and with respect to what types of data can be generated (Chapter 3).

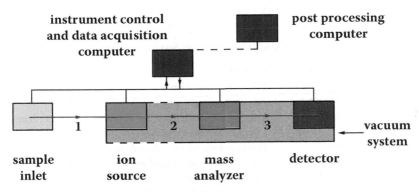

1. Vaporized sample enters the ion source
2. Unresolved ion beam is accelerated to the analyzer
   where the ions are separated either 'in space' or 'in time'
3. Ions with resolved *m/z* travel to the detector

Mass spectrometers are composed of a sample inlet, ion source,
   mass analyzer, detector, vacuum system, and computer.
   Sometimes these components are combined, e.g., the inlet and
   ion source, or the analyzer and detector.
While multiple types of each component can be mixed and
   matched, e.g., different ion sources with different analyzers,
   the ionization, analysis, detection sequence must be maintained.
Computers control instrument operation, data acquisition, and
   data processing.

FIGURE 1.2   Block diagram of a mass spectrometer.

The process of choosing the type of mass spectrometer (including the sample inlet system) and operational conditions (experiments) to be undertaken to analyze a sample is based on several factors, including the molecular mass and polarity of the analyte(s) and the complexity of the sample mixture. The information sought, be it structural or quantitative (or both), also influences the selection of instrument and experimental setups.

What is mass? In everyday usage, *mass* is usually taken to mean weight and ignores any reference to gravitation. In physics (without invoking major discussions), mass refers to the quantity of matter that a body contains. Mass is measured in kilograms in the International System of Units (SI). Mass spectrometry tends to use a hybrid system of units, for instance, referring to molecular weight when molecular mass would be preferable and using dalton (Da) as the unit of mass. Dalton is accepted by the International Union of Pure and Applied Chemists (IUPAC) but is not an SI unit.

What are ions? *Ions* are atoms, molecules, or fragments of molecules that carry one or more positive or negative electrical charges. An ion is created when the number of protons (i.e., positive charges) in the nucleus of a molecule is no longer

---

TABLE 1.1
Types of Ions Commonly Observed in Mass Spectrometry

Ion	Example	Description
Cation	$[M + H]^+$	Positively charged ion
Anion	$[M - H]^-$	Negatively charged ion
Molecular ion	$[M]^{+\bullet}$	Positive radical ion, formed in electron ionization, by removal of an electron from a molecule
	$[M]^-$	Negative molecular ion, formed in electron capture negative ionization, by addition of an electron to a molecule
Protonated molecule	$[M + H]^+$	Positive ion, formed in chemical ionization and atmospheric pressure ionization, etc. (i.e., soft ionization methods), by adding a proton to a molecule
Deprotonated molecule	$[M - H]^-$	Negative ion, formed in atmospheric pressure ionization, by removal of a proton
Adduct ion	$[M + Na]^+$ $[M + Cl]^-$ $[M + NH_4]^+$	Any ion formed by the addition of an ionizing species to a molecule, e.g.: • Alkali metals in positive electrospray • Cl⁻, CH_3COO^- in negative electrospray • Addition of products derived from the reagent gas in chemical ionization
Multiply charged ion	$[M + nH]^{n+}$	Product formed by adding multiple protons, typically to a peptide or protein, during electrospray ionization
Fragment ion	$[X]^+$	Product observed when the energy added to a molecule during ionization results in the breaking of a chemical bond(s) and the generation of a charged species
Isotope ion	^{13}C, ^{29}Si, ^{30}Si, ^{34}S, ^{37}Cl, ^{81}Br	Ion resulting from the inclusion of elements for which there are alternate stable forms (isotopes)

balanced by the number of negatively charged electrons present, e.g., by the addition of a proton or removal of an electron that creates, in both cases, a species with a +1 charge. In mass spectrometry, analytes must be converted into ions (ionized), and must be in the gas phase, so that they can be manipulated by electrical or magnetic fields in order to determine their masses. There are a number of significantly different methods to make ions, generically classified as hard and soft, that lead to the formation of a variety of ion types (Table 1.1).

Hard ionization, in particular electron ionization (Section 2.2.1), imparts more energy than that required for the formation of molecular ions. The distribution of the excess energy frequently causes the dissociation of bonds within the analyte molecules, resulting in extensive fragmentation.

In soft ionization, e.g., chemical and electrospray ionization, the energy associated with the added charge is usually only sufficient to generate adducted ions, such as $[M + H]^+$, i.e., M + proton (Section 2.2.2), but not to cause fragmentation. Note that the ions formed are protonated molecules and not protonated molecular ions, as the latter would have two charges, i.e., the already ionized molecular ion and the proton that is itself an ion.

Mass spectrometry has traditionally dealt mainly with positively charged ions, and this is still the ion polarity formed most commonly. However, instruments are now designed to handle both positive (cationic) and negative (anionic) ions.

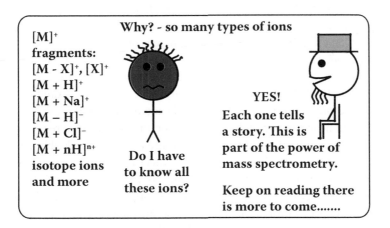

Why? - so many types of ions

$[M]^+$
fragments:
$[M - X]^+$, $[X]^+$
$[M + H]^+$
$[M + Na]^+$
$[M - H]^-$
$[M + Cl]^-$
$[M + nH]^{n+}$
isotope ions
and more

Do I have to know all these ions?

YES!
Each one tells a story. This is part of the power of mass spectrometry.

Keep on reading there is more to come.......

What is mass-to-charge (m/z) ratio? The mass-to-charge ratio of an ion is the number obtained by dividing the mass of the ion (m) by the number of electrical charges (z) acquired by the sample during the ionization process. The m/z of an ion is a dimensionless number: m and z are always written in italics. Because of the lack of dimensions, no equals sign should be used when specifying an ion, e.g., m/z 201 and not $m/z = 201$. The use of the term *Thomson* (Th) has been proposed for m/z, but it is used infrequently.

Mass spectrometers are m/z analyzers; they do not directly measure mass.

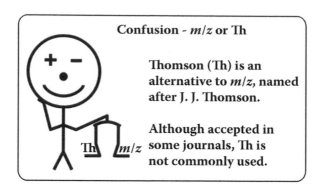

What is m in m/z? The scales of atomic masses are based upon an agreed standard. The currently accepted standard (IUPAC) is the *unified atomic mass unit* (u), a dimensionless number, defined so that a single carbon atom (the carbon-12 isotope that contains 6 protons and 6 neutrons) has a mass of *exactly* 12 u (1 u = 1.66×10^{-27} kg.) The unit u is also called a *dalton* (Da). Dalton is used in preference to atomic mass unit (amu), with the latter considered archaic. The Da is convenient for expressing the masses of both atoms and molecules, e.g., argon (Ar) = 40 Da, methane (CH_4) = 16 Da, and cholesterol ($C_{27}H_{46}O$) = 386 Da.

What is z in m/z? The *electrical charge* (positive or negative) present on an ion is represented by z. A single charge is the electromagnetic force associated with a proton (+1) or an electron (–1). In most cases there is only one charge on an ion; thus, the measured m/z value is equivalent to the mass of the ion ($z = 1$). When there are additional charges on the ion, the measured m/z value will be the quotient of the mass of the ion divided by the number of charges, i.e., $z > 1$. For example, the addition of a single proton ($m = 1$, $z = 1$) to a molecule of mass 399 produces an ion of (399 + 1)/1, and hence an m/z of 400. If, however, this molecule were to acquire two charges, there would be the addition of two protons yielding a doubly charged ion ($m = 2$, $z = 2$) with m/z 200.5, i.e., (399 + 2)/2. Electrospray ionization (ESI), which is one of the most important current techniques for the ionization of large, complex biomolecules, produces an *envelope* (array) of multiply charged ions from a single analyte with z up to 75 (or more), depending on the molecular mass of the analyte. The signals from these ions create an envelope that can be *deconvoluted*, using specialized software, to account for all the different charge states present, and thereby derive the mass of the analyte (Section 2.2.2.2).

What is a mass spectrum? A mass spectrum is a two-dimensional representation, graphical or tabular, of the distribution and intensity (abundance) of the ions introduced into the mass analyzer and then separated and recorded according to their m/z values (Figure 1.3). It is important to remember that the parameter plotted on the x-axis is the mass/charge ratio (m/z) rather than, as often thought, just the mass (m). Figure 1.3 shows four basic features (regions) of mass spectra:

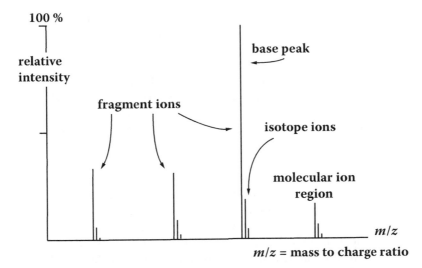

m/z = mass to charge ratio

Energy is added to molecules during ionization. The distribution of the energy may result in the breaking of chemical bonds and, consequently, in fragment ion formation. The fragmentation may be so extensive that no molecular ion is observed.

The form of the molecular ion depends on the mode of ionization and can include $[M]^+$, $[M + H]^+$ and other adduct ions, e.g., $[M + Na]^+$.

The base peak represents the most stable ion resulting from the ionization process and is, therefore, the most intense (abundant) peak in the spectrum. The intensities of all other ions are usually normalized with respect to the base peak.

Ions, normally of lesser intensity and to the right of each molecular/fragment ion, generally represent isotopic species. Typically, but not always, isotope ions reflect the presence of carbon-13 (^{13}C).

FIGURE 1.3 Basic features of mass spectra.

1. The *molecular mass (weight)* of the analyte is often (but not always) represented by the peak with the highest *m/z* value, i.e., the one furthest to the right on the x-axis. This peak corresponds to the *molecular ion*, $[M]^{+\bullet}$, or a directly related species, such as the *protonated molecule*, $[M + H]^+$.

2. The *base peak* represents the most stable ion resulting from the ionization process, thus it is the most intense (abundant) peak in the spectrum. The intensities of all other ions are usually normalized with respect to the base peak.

3. *Fragment* ions may appear at various masses, but their masses are always lower than that of the molecular mass, and reflect the fact that the amount of energy added to molecules during ionization may lead to the decomposition of the molecular ion as well as further fragmentation of any subsequently formed ion.

4. *Isotope* peaks are observed if the analyte includes elements with isotopic forms. The most commonly observed isotope peaks are those due to the ^{13}C isotope of carbon.

Mass spectra may be presented in three forms: (1) The *tabular* form comprises the *m/z* and intensities of the signals produced by ions arriving at the detector. Spectral data are stored in tabular form; however, this format is used infrequently because of the size and inconvenience of the data files and the lack of a visual entity. (2) The *analog* (profile) form is where each peak has a height and a width and is displayed as a continuous outline. (3) The *digital* (bar graph) form is where each peak is represented by a simple vertical line (Figure 1.4). As shown in the figure, data are acquired initially by the data system in a digital (tabular) format that is subsequently plotted in the analog (profile) form. This is usually converted to a simplified bar graph output composed of vertical lines drawn through the *centroid* of each peak profile with heights that reflect the signal intensities of the ions. Note that it is also possible to convert incoming data to the simplified bar graph form, thereby saving disc space, but there is a consequent loss of information, such as peak shape.

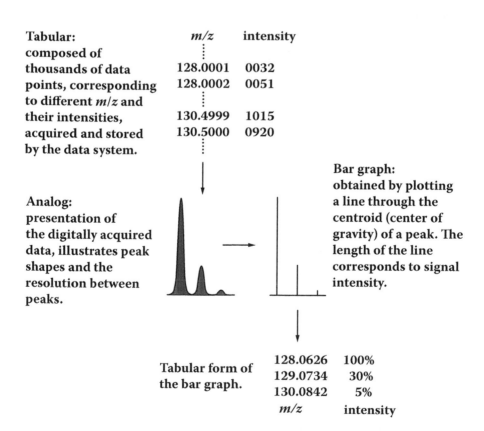

Tabular: composed of thousands of data points, corresponding to different *m/z* and their intensities, acquired and stored by the data system.

m/z	intensity
⋮	
128.0001	0032
128.0002	0051
⋮	
130.4999	1015
130.5000	0920
⋮	

Analog: presentation of the digitally acquired data, illustrates peak shapes and the resolution between peaks.

Bar graph: obtained by plotting a line through the centroid (center of gravity) of a peak. The length of the line corresponds to signal intensity.

Tabular form of the bar graph.

m/z	intensity
128.0626	100%
129.0734	30%
130.0842	5%

FIGURE 1.4   Tabular and graphical forms of mass spectra.

Ion intensities are derived from the minute currents created as the ions arrive at the detector. These currents are expressed and stored in arbitrary units that represent the areas and heights of the peaks, and spectra may be presented in this manner. However, it is more common to plot ion intensity data as *relative intensities*, normalized with respect to the *base peak*, i.e., the peak of highest intensity (within a selected mass range), be that the molecular or a fragment ion.

What are peaks? A *peak* is defined as the time during data acquisition where the concentration (abundance) of a given ion (specific *m/z*) is larger than the background electrical noise. (Note that in chromatography the term *peak* refers to the detection of a separated component, from an injected mixture, eluting from a chromatographic column).

What are isotope peaks? The atoms of elements are composed of protons, electrons, and neutrons. An element is defined by the number of protons present in the nucleus. However, the number of neutrons may vary over a small range, yielding products known as *isotopes*. For example, there are three isotopes of carbon (which has six protons) incorporating six, seven, or eight neutrons yielding ^{12}C, ^{13}C, and ^{14}C, respectively. The presence of the ^{13}C isotope, which occurs with a natural abundance of ~1.1% per carbon atom, results in the isotope peaks that are commonly found on the right-hand side of all ions in mass spectra (Figures 1.3 and 1.5). Because the abundance of ^{14}C is less than 0.001%, this isotope is not observed using conventional mass spectrometers. Many elements display distinctive isotope patterns, e.g., in compounds containing one atom of chlorine (^{35}Cl and ^{37}Cl, ratio $\approx$ 3:1) or bromine (^{79}Br and ^{81}Br, ratio $\approx$ 1:1); the isotope peaks occur as doublets two mass units apart. The natural abundances of the isotopes of these halogens lead to the appearance of unique, and diagnostic, isotope patterns (Figure 1.5). For many metals, e.g., platinum, there are more complex isotope *clusters* (Figure 1.5). Judicious evaluation of isotope patterns can often assist in the determination of elemental compositions.

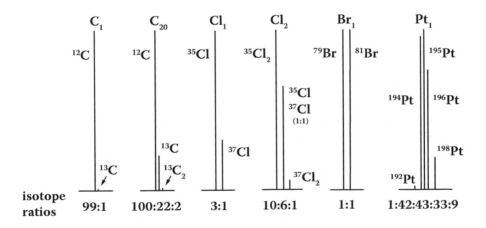

FIGURE 1.5 Examples of isotope patterns.

The frequency of the ^{13}C isotope is of major importance in the mass spectra of biopolymers, as the molecular masses of the compounds reflect the contribution from this isotope. For example, a protein containing 1,000 carbon atoms contains, on average, 10.7 ^{13}C atoms and has a molecular (average) mass that is 10.7 mass units larger than that an ion from a molecule that contains only ^{12}C.

What are the differences between nominal mass, exact mass, accurate mass, mono-isotopic mass, molecular mass, and isobaric mass? The *nominal* mass of a compound, ion, or fragment is calculated using the masses of the elements rounded to the nearest whole number (e.g., C = 12, H = 1, O = 16). For example, cholesterol with an empirical formula $C_{27}H_{46}O$ has a nominal molecular mass of 386 Da. It is common to see molecular mass referred to as molecular weight (MW).

The *exact* mass is a calculated value, using the masses of the isotopes of the elements based on the unified atomic mass unit scale (C = 12.0000, H = 1.007825, O = 15.9949); e.g., the exact molecular mass of cholesterol is 386.3549 Da.

Accurate (ionic) mass is the experimentally determined mass of an ion. The value obtained is a function of the design, resolution, tuning, and calibration of the mass spectrometer.

Mono-isotopic mass is the calculated mass of a compound based on the accurate masses of the most common stable isotopes of the elements present. For the elements commonly found in organic molecules the mono-isotopic mass is composed of the isotopes of lowest mass, e.g., 12.0000 for carbon and 15.9949 for oxygen. The mono-isotopic mass is usually the most relevant measure of mass, because of the ability of instruments to observe the masses of the individual isotopes.

Molecular mass is the average mass of a compound obtained when an accounting is made for all isotopes of the elements present based on their relative abundances, e.g., C = 12.0108 Da, O = 15.9994 Da. If the instrument used cannot resolve the individual isotopes, the observed peaks include all isotopes present. Molecular mass is sometimes referred to as *average* mass. For cholesterol the molecular (average) mass is 386.6616 Da. (Note the difference between this number and the mono-isotopic exact mass value, 386.3549 Da, as described above.) Molecular mass does have a dimension as it is an absolute value unit, based on 1/12 of the mass of the ^{12}C isotope (in IUPAC units), i.e., 1.6605×10^{-27} kg. If, however, the mass of an analyte is considered as a ratio with respect to the mass of ^{12}C, then the dimensions cancel out and the resulting dimensionless number is the *relative molecular mass*. These two terms are equivalent in everyday usage.

Molecular mass is particularly relevant when (bio)polymers are analyzed because mass resolution is frequently insufficient to observe individual isotopes. For these compounds the ion observed is a composite made up of the average of all isotopes present. The prevalence of carbon and the relatively high frequency of the ^{13}C isotope (1.1%) means that the ^{13}C isotope has an important influence on the masses observed in the spectra of biopolymers. For example, the molecular mass of a 20 kDa protein is approximately 10 mass units higher than that of its mono-isotopic mass.

Isobaric mass describes empirical formulae that have the same nominal mass but different exact masses; e.g., CO and N_2 are isobaric at mass 28 but have exact mass values of 27.9949 and 28.0062 Da, respectively.

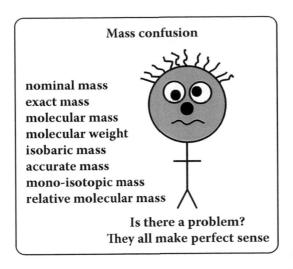

What is ionic mass? The *ionic* mass of an ion takes into account the mass of an electron (0.000548 Da ≡ 0.548 mDa; mDa is also referred to as millimass units, mmu) that is removed or added during the formation of the ion (remember that a proton is a hydrogen atom, ^1H isotope, after removal of an electron). This small mass effect is frequently ignored when masses are determined experimentally and compared with calculated values. However, as the accuracy and precision of accurate mass measurements have improved, the mass attributable to an electron has become of relevance to empirical formula determinations. The mass of an electron represents ~1 part per million at 500 Da (~0.5 mDa), which is a significant error in *m/z* determinations for instruments designed to obtain mass accuracies at the mDa level or less.

1.3 COMPONENTS OF INSTRUMENTS AND THEIR FUNCTIONS

1.3.1 Sample Introduction

Mass spectrometers operate at high vacuum (Section 2.5), thus they can only analyze samples that are in the vapor state. Equally importantly, the neutral analyte molecules must be converted into ions. The functions of sample introduction systems are to produce vapors from samples (or reduce the pressure of gaseous samples) and to introduce a sufficient quantity of the sample into the ion source in such a way that its composition represents that of the original sample. It is important to note that the concept of sample introduction followed by ionization has changed with the development of recent techniques where the sample introduction and ionization process occur simultaneously. These techniques include atmospheric pressure ionization (API), particularly electrospray ionization (ESI) and atmospheric pressure chemical ionization (APCI).

Single compounds or simple mixtures can be introduced directly using a variety of methods. These include the temperature-controlled *solids introduction probe* (also called a direct insertion probe) that is inserted through a *vacuum lock*, and a *heated volume* from which evaporated samples enter the ion source through a control valve. The latter is often used in electron and chemical ionization instruments to introduce the compounds used to calibrate the mass scale.

The requirement that samples be in the vapor phase used to be a limiting factor in the utility of mass spectrometry because it was frequently necessary to add heat to vaporize samples. Successful vaporization depends on the polarity and molecular mass of the analyte. For small molecules (<1 kDa) polarity is the most important factor, particularly the number of primary amino ($-NH_2$), hydroxyl ($-OH$), and acidic ($-COOH$) groups present. When there are two or three of these groups present, heating will frequently destroy compounds before they vaporize, even at reduced pressures. Chemical derivatization, e.g., the conversion of an alcohol to its trimethylsilyl ether or the acetylation of an acid group, can be used to reduce the polarity and increase the volatility of analytes, thereby improving the chances of obtaining useful spectra.

When the analyte is >1 kDa, heating is likely to decompose the molecule irrespective of the presence or absence of specific groups. The development of API and MALDI, where vapor formation and ionization occur simultaneously, has resolved most problems of sample decomposition.

Mass spectrometers with simple sample introduction systems are not effective when analyzing complex mixtures, although multianalyzer systems can address this problem to some extent (Section 3.3.2). Samples of biological or environmental origin usually require chromatographic separation with subsequent, sequential introduction of the constituent species into the ion source. Gas chromatography (GC), liquid chromatography (variously abbreviated as LC, HPLC, and UPLC), and capillary electrophoresis (CE) can be interfaced directly with mass spectrometers (Section 2.1.2).

Gas chromatography requires that samples remain stable when volatilized in a stream of helium, first in the injector used to introduce the analytes onto the GC column, and then during the time that the sample components traverse the column as it is heated progressively inside an oven. The requirement for volatility means that only nonpolar or moderately polar compounds can be analyzed by GC-MS. For capillary GC columns the flow rate of the gas that moves compounds through the column (the *carrier gas*) is low, about 1 ml/min. Such a quantity of gas can be introduced into the ion source directly, without compromising the vacuum in the instrument. This simplifies the interfacing of GC with MS so that a heated interface tube can be used to link the two instruments and through which the GC column is run so that it abuts the ion source.

The major difficulties associated with LC-MS are not only the vaporization and ionization of polar compounds without their decomposition, but also the volume of gas generated as the LC elution solvent transitions from liquid to vapor phase. These issues were resolved with the development of the API methods that have enabled the direct interfacing of reversed-phase LC columns with MS. The coupling of these techniques has expanded the utility of mass spectrometry into the analysis of biologically relevant molecules, most of which are highly polar. In addition, LC is compatible with compounds of high molecular mass, including peptides, proteins, and other biopolymers. Capillary electrophoresis (CE), a specific form of LC, can also

TABLE 1.2

Advantages and Disadvantages of Various Sample Introduction Methods

Advantages	Disadvantages
Solids Probe	
Compatible with nonpolar compounds up to ~1 kDa	Must be inserted through a vacuum lock
	Single compounds or very limited mixtures
	Ineffective for polar compounds that decompose when heated
Batch Inlet	
Continuous controlled flow of sample for process monitoring	Sample must volatilized readily
Introduction of calibration compounds	
Gas Chromatography (GC)	
Efficient separation	Analytes must be nonpolar
Convenient interfacing	Molecular mass limited to ~600 Da depending on structure
Liquid Chromatography (LC)	
Can be used with polar compounds, including biopolymers with molecular mass 100 kDa and higher	Separations not as efficient as GC
	Eluting liquids produce large volumes of vapor
Capillary Electrophoresis (CE)	
Efficient separations of large biological compounds	Very low flow rates make interfacing to MS difficult

be interfaced with an API source. A summary of the advantages and disadvantages of various sample introduction methods is given in Table 1.2.

1.3.2 Ion Sources

Molecules are electrically neutral, so they cannot be manipulated using electrical or magnetic fields. By analogy, think of trying to use a magnet to attract a piece of brass (analogous to a neutral molecule). Nothing will happen. If a second magnet is attached to the brass (equivalent to ionization), it will be possible to attract or repel the object depending on which poles of the magnets are facing each other: like poles repel, unlike poles attract. In a similar way, electrical charges, and thus charged species (ions), can be manipulated in electrical (and also magnetic) fields.

The fact that the focusing, separation, and detection processes in a mass spectrometer require charged species means that the next event, after sample introduction, must be the conversion of the molecules into ions (second block of Figure 1.2). Historically, most ions studied were positively charged, and this remains the polarity used most frequently, but modern instruments handle both positive (cationic) and negative (anionic) ions.

The process of ionization adds energy to the analyte molecules. Controlling the amount of energy affects the ionic products generated and, consequently, the types of molecular species or fragments produced. There are a wide variety of ways to form ions both inside and outside the mass spectrometer. The method of ionization determines the amount of energy added to the analyte. Until the 1990s, all ionization (primarily electron and chemical ionization) occurred in ion sources located within the vacuum chamber of the mass spectrometer. The ionizing "agents" included electrons, other ions, electric fields, and lasers. In recent years, methods have been developed to effect ionization outside the vacuum chamber, at atmospheric pressure. These methods have significantly increased the diversity and range of applications of mass spectrometry.

Electron ionization. *Electron ionization* (EI) was the first widely used method, and it is still used, most often in GC-MS. In EI, analytes in the gas phase are bombarded with energetic (70 eV) electrons obtained from a heated filament located inside the vacuum chamber. The energy imparted to the analytes is in considerable excess over that required to ionize organic species (typically ~10 eV)—thus the term *hard ionization*. The bombardment removes an electron from the sample molecules. The initial product is a positively charged molecular ion, a radical cation, $[M]^{+\bullet}$. Although "M plus dot" ($[M]^{+\bullet}$) is the correct designation for a radical molecular ion, the more commonly used $[M]^{+}$ will be the format used here. The excess energy imparted during EI is distributed among the bonds of the ions formed and often leads to the dissociation of one or more bonds and the ejection of radical or neutral species. The resulting fragment ions will have lower masses, and lower energies (because of the energy consumed during the fragmentation process), than the molecular ion. The fragmentation may be so extensive that no molecular ion can be observed. This can be frustrating because the molecular mass of a compound is often the most useful piece of information provided by a mass spectrometer. However, the fragmentation is generally predictable and provides information on the structure of the analyte. This predictability has made it possible to construct libraries of spectra that can be searched (by computer) to provide tentative identifications of unknown analytes.

A major disadvantage of EI is that it is limited to molecules with molecular masses of <1 kDa.

Chemical ionization. The need to determine the molecular masses of analytes led to the search for soft ionization methods that reduce the amount of energy delivered to the molecules in the ionization process and, consequently, limit fragmentation. *Chemical ionization* (CI) was the first of the soft methods. In CI, electrons from a heated filament react with a reagent gas (e.g., methane) to form ions that, in turn, interact with the analyte onto which a proton (i.e., a hydrogen nucleus) is transferred, thus producing protonated molecules, $[M + H]^{+}$. This type of ion is more stable than those formed by EI because less energy is transferred during the CI process. The fact that $[M + H]^{+}$ are the predominant species in CI, and that there is little fragmentation, favors the use of CI in quantification, most often involving GC-MS.

Similarly to EI, the upper mass limit of CI is ~1 kDa.

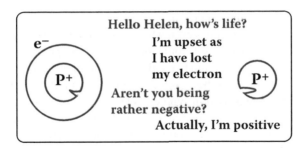

Atmospheric pressure ionization (API). The need to analyze polar compounds and the necessity to interface LC with MS led to the development of techniques where the ionization occurs at *atmospheric pressure* outside the vacuum chamber, and the resulting ions are transferred directly into the mass analyzer. Electrospray ionization (ESI) is the most successful of the API methods because of the range of molecular masses to which it can be applied, from small molecules to proteins. Other API methods include atmospheric pressure chemical ionization (APCI) and atmospheric pressure photo-ionization (APPI), and also the recently developed *surface ionization* methods such as desorption electrospray ionization (DESI) and direct analysis in real time (DART) (see below and Sections 2.2.2 and 2.2.3).

Atmospheric pressure chemical ionization (APCI). APCI is the equivalent of CI conducted at atmospheric pressure. Here both the analyte and the LC solvent are evaporated and the vast excess of solvent vapor is used to produce reagent ions by using a *corona discharge* to form an ionizing plasma. The discharge emanates from a pin that replaces the filament used in EI and CI that would burn out immediately at atmospheric pressure. The reagent ions from the solvent are the equivalent of those obtained from the reagent gas used in CI. Ionization occurs through proton transfer from the reagent ions to the analyte-forming protonated molecules, $[M + H]^+$. APCI is often the method of choice for the quantification of small molecules, as it is a chemical ionization process that provides a more uniform analyte response than ESI. As with EI and CI, the upper mass limit of APCI is ~1 kDa.

Electrospray ionization (ESI). This method is based on the effect of placing a high voltage (~2–5 kV) on a conductive solvent in which the analyte is dissolved and spraying the solution through a hollow needle. Ions of the charge opposite to that of the applied voltage are stripped from the solution. The charged, nebulized droplets are then evaporated continuously, first at atmospheric pressure and then as they pass into a series of progressively evacuated chambers that form the connection between the source and the mass analyzer. The charge density of the droplets increases as they evaporate, and eventually a point is reached where the integrity of the droplets can no longer be maintained and ions, with the same polarity as the applied voltage, are released into the vapor phase. ESI is compatible with polar compounds; it is liquid based and requires a conductive solvent. These attributes make this ionization process compatible with reversed-phase LC and, consequently, with the very large class of compounds that are typically analyzed by this form of chromatography. ESI has made possible the expansion of mass spectrometry into biological analyses where most compounds are water soluble and polar. One disadvantage of ESI is that in mixtures it favors the most polar compounds even when those analytes are minor constituents.

Ions formed using this very soft ionization method are predominantly $[M + H]^+$ or $[M + \text{alkali metal}]^+$, with sodium being the most commonly observed alkali metal. The major source of sodium is the biological samples themselves, but there are other sources, including reagents used in synthetic chemistry and even glassware.

A unique, and very important, feature of ESI is that as the molecular mass of an analyte increases, the ionization process often leads to the formation of multiply charged ions, of the form $[M + nH]^{n+}$. The consequence of adding multiple protons to the analytes is that an *envelope* of charged species is formed with each member separated from the next one by a single charge. The generation of multiply charged ions has facilitated the use of mass spectrometers for the analysis of *intact* biopolymers, particularly proteins. An example of an ESI spectrum of a protein, myoglobin, is given in Figure 1.6a. The progressively increasing spacing between the peaks is characteristic of multiple charging. The relationship between the masses of the peaks can be expressed by a set of simultaneous equations the solution of which provides the molecular mass of the protein. The accounting for the multiple charges that enables the calculation of a theoretical spectrum of the neutral molecule (zero charge state) is called *deconvolution (transform)*. The result of deconvolution for myoglobin is shown in Figure 1.6b. Because the *m/z* value corresponding to each charge state can be measured accurately, the derived mass of the protein is also accurate, often within a mass unit or two. An important advantage of the multiple charging in ESI is the possibility to analyze proteins and other compounds with high molecular mass, using inexpensive instruments, such as quadrupoles, that have mass ranges that are much less than the mass of the analyte.

Matrix-assisted laser desorption/ionization (MALDI). This is another ionization method for the analysis of large molecules such as peptides, proteins, and nucleic acids, as well as some synthetic polymers. In MALDI, the analyte is first co-crystallized with an excess of a *matrix*, e.g., sinapinic acid or dihydroxybenzoic acid, that has a constituent aromatic component able to absorb photons from a UV laser beam. When the dried analyte:matrix mixture is exposed (inside the vacuum chamber) to a sudden input of energy from a laser pulse the matrix evaporates, essentially instantaneously, carrying with it the analyte molecules. The matrix forms reagent ions that protonate the analytes. The selection of the matrix is critical as different compound classes exhibit substantial, matrix-dependent differences in ionization efficiency. The MALDI matrix should not be confused with the alternative use of the term matrix that is used to denote the medium in which biological and/or environmental components are presented, e.g., blood plasma, urine, sediment.

In contrast to ESI, the majority of ions generated by MALDI carry a single charge; thus, the analyzer must have an upper mass limit sufficiently high to deal with ions where $z = 1$. Time-of-flight (TOF) analyzers are almost always used with MALDI (Sections 2.2.4 and 2.3.3). MALDI-TOF instruments provide rapid (and routine) analysis of biologically important molecules with molecular masses up to 350 kDa. MALDI spectra are easier to interpret than ESI spectra. MALDI spectra consist of single peaks with limited mass accuracy that broaden rapidly, becoming several

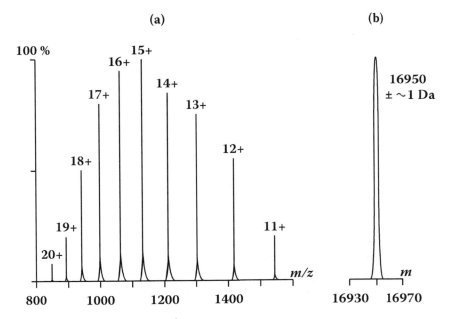

(a) **Envelope of multiply charged ions (n+ = charge state)**
   showing the progressively increasing spacing that indicates
   the presence of a protein (or other biopolymer).
(b) **Computerized deconvolution (transform) of the multiple**
   charge state envelope provides the molecular mass of the
   neutral protein ($z = 0$).

FIGURE 1.6 Electrospray ionization (ESI) mass spectrum and deconvoluted data for apo-myoglobin.

hundred mass units wide near the x-axis. The simplicity as well as the drawbacks of MALDI spectra of proteins is illustrated for myoglobin in Figure 1.7.

Comparison of ionization techniques. Table 1.3 gives an overview of the advantages and disadvantages of the various ionization methods. Each method, and some others, is revisited in more detail in Sections 2.2, 3.2.3, and 3.2.4.

1.3.3 MASS ANALYZERS

The role of the *mass analyzer* is to separate ions according to their *m/z* values and to focus and transfer these ions onto a detector, or into a collision cell in multianalyzer instruments (see later). The mass analyzer is the heart of all mass spectrometers (block 3 in Figure 1.2). The choice of which analyzer to use is critical as it affects multiple aspects of the data generated, including mass resolution, mass measurement accuracy, and available dynamic range. There are several types of analyzer: quadrupole (Q), ion trap (quadrupole (QIT) or linear (LIT) ion trap), time-of-flight (TOF),

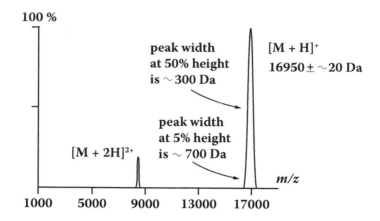

The mass observed is for apo-myoglobin; the heme group is lost.
Spectrum consists primarily of [M + H]⁺ ions; simpler than ESI.
Peaks are broad because resolution is limited.
Accuracy of mass determination for biopolymers is ±0.1%.

FIGURE 1.7   MALDI-TOF spectrum of myoglobin.

magnetic sector (B), orbitrap, and ion cyclotron resonance (ICR) cell. The last two are the so-called "Fourier transform" (FT) instruments.

Analyzers separate ions in time or in space. In the Q, QIT, LIT, and B analyzers operational parameters must be changed progressively over time (scanning) in order to obtain spectra. Such scanning instruments are *in-time* systems, because the ions are collected successively over a period of time (milliseconds to seconds). Scanning does increase the versatility of an instrument, but the cost is reduced sensitivity when collecting spectra because of a *duty cycle* issue, i.e., the limited amount of time spent analyzing a given mass during the collection of an entire spectrum. In contrast, the *in-space* analyzers, TOF and FT, are nonscanning with all ions detected simultaneously (FT) or almost simultaneously (microsecond scale) in TOF systems. The in-space strategy improves sensitivity when collecting full spectra, but prevents the use of other important techniques, such as selected reaction monitoring (Section 3.3.3.1).

Almost any type of source can be coupled with any type of analyzer because ions are usually produced in locations separate from where they are analyzed. MALDI is an exception, as it cannot be coupled with quadrupole analyzers because of their limited mass range.

Analyzers may be combined, almost in a mix-and-match manner, into sophisticated *tandem* instruments, e.g., QqQ, QTOF, and LIT-FT-ICR, which are capable of providing significantly more information than single analyzer systems. Tandem analyzers are also referred to as mass spectrometer/mass spectrometer (MS/MS) instruments. These terms apply to the combination of two (or more) analyzers to attain improved detection limits or to provide hitherto unavailable structural information. *Tandem* often implies that the instruments utilize like analyzers, such as in

TABLE 1.3
Advantages and Disadvantages of Various Ion Sources

Advantages	Disadvantages
Electron Ionization (EI)	
Well understood	Analyte <1 kDa
Reproducible across multiple instrument types	Analyte must be volatile
Structural information from fragments	Spectra often complex
Searchable computer libraries	Spectra may lack molecular ion
Chemical Ionization (CI)	
Usually provides molecular mass information	Analyte <1 kDa
	Analyte must be volatile
	Spectra depend on the choice of reagent gas
Atmospheric Pressure Chemical Ionization (APCI)	
Ionization is more uniform than ESI	Analyte <1 kDa
Compatible with LC	
Electrospray (ESI)	
Effectively ionizes polar (nonvolatile) compounds, <100 Da to >10^6 Da	Efficiency of ionization depends on polarity
Soft ionization often gives only molecular mass-derived ions, e.g., $[M + H]^+$, $[M + Na]^+$, $[M + nH]^{n+}$	
Compatible with LC	
Applicable to noncovalent interactions	
Matrix-Assisted Laser Desorption/Ionization (MALDI)	
Mass range 500 Da to >350 kDa	Sample preparation affects ionization
	Off-line technique

the triple quadrupole (QqQ) and TOF-TOF systems, whereas *hybrid* instruments combine dissimilar analyzers, for instance, in the QTOF or LIT-orbitrap. MS/MS analyses are particularly important when combined with soft ionization techniques because these methods only provide molecular mass data. The need for structural information led to specific strategies for fragmenting ions that are subsequently analyzed in the second analyzer. These methodologies are discussed in Section 2.3.7 and in several parts of Chapter 3.

When choosing an analyzer, or combination of analyzers, the type of data sought, i.e., structural or quantitative, the required degree of resolution, and mass measurement accuracy should all be considered.

You told me that everything would become clear. All I see is a ridiculous alphabet soup.

GC-MS, LC-MS, MS/MS, EI, CI, ESI, MALDI, CID, ETD, ECD, QqQ, FT-ICR. OK, they do sound like government agencies. C'est la vie. Just keep on reading...

Quadrupole analyzers (Q). In these transmission type analyzers ions are separated according to their m/z in an electric field generated in the space within a set of four rods arranged precisely in a square, with opposite pairs of rods connected electrically. The rods are typically ~20 cm long by 0.5–1.0 cm in diameter. Combinations of rf voltages (180° out of phase) and dc voltages (both positive and negative) placed on the rods generate fields that enable transmission of ions from one end of the analyzer to the other, with specific voltages being required for the transmission of ions for each m/z value. Spectra are collected by changing the rod voltages sequentially over time. Alternatively, just one mass can be monitored if a specific set of voltages is placed on the rods. The latter is the method of choice for quantification. The mass range of these analyzers for singly charged ions is up to 4 kDa. A limiting factor of quadrupole analyzers is that only unit resolution is available across their entire mass range; e.g., 200 is resolved from 201, and 1,000 from 1,001.

An important configuration of quadrupoles is the triple quadrupole (QqQ), in which there are two analytical quadrupoles (Q) separated by a transmission quadrupole (q). While the predominant use of the QqQ is in quantification, this very versatile format has several scanning modes that enable multiple MS/MS approaches to obtain structural information (Section 3.3.3.1). Extensions of the quadrupole technology are the quadrupole ion trap (QIT) and the more recent linear ion trap (LIT) that has higher ion capacity. The resolutions of these ion traps are similar to those of single quadrupoles. However, an advantage of the traps is the ability to store and manipulate ions prior to their detection, thus enabling MS/MS experiments (Section 2.3.2).

Time-of-flight (TOF) analyzers. TOF analyzers are inherently simple systems that measure the time it takes for an ion of a given m/z to travel from the source to the detector. Packets of ions are pulsed into the analyzer (using an accelerating voltage) where they travel through an evacuated tube (typically 1–2 m long) to a detector. The ions are separated according to their momenta, and therefore their m/z ratios; lighter ions travel faster than heavier ions. The development of specialized techniques, including delayed extraction and orthogonal ion injection, as well as reflectrons (all discussed in Chapter 2), has contributed to improvements in resolution, with up to 60,000 currently available.

TOF analyzers are nonscanning because they are based on separation of the ions within the space of the flight tube and do not require the changing of electronic parameters to allow for the passage and detection of ions with different m/z. Accurate mass data can be obtained for all transmitted masses in a fraction of a second. The rate of data acquisition is of particular relevance given that sample introduction is usually via chromatography. The faster a compound elutes from a column (a desirable aim to reduce per sample analysis time), the more rapid must be the subsequent data collection and analysis. The capabilities of high-resolution and accurate mass measurement and the rapid rate of data acquisition have led to the widespread use of TOF instruments. Tandem/hybrid systems using TOF analyzers include the TOF/TOF and QTOF formats (Sections 2.3.7.1.2 and 2.3.7.2.1).

Fourier transform (FT) analyzers. The FT instruments have the highest available resolution. Currently there are two types, the orbitraps and the ion cyclotron resonance (ICR) systems. Mass resolution in orbitraps can reach 250,000, while ICR systems can have resolutions of >3,000,000. In these analyzers ions oscillate/rotate within a cell and are detected by recording the electrical current that the passage of ions induces in the surfaces of the cell. The resolution attainable is inversely proportional to both the mass of the analyte ion and the time required to acquire the data. A consequence of this proportionality is that the highest resolutions cannot be obtained on the chromatographic timescale where multiple spectra must be collected to enable the characterization of peaks that are only a few seconds wide. LIT/FT combinations are the most common form of MS/MS systems that utilize orbitrap and ICR analyzers (Sections 2.3.4 and 2.3.7.2.2).

Comparison of mass analyzers. An overview of the most important performance parameters of the various analyzers is provided in Table 1.4.

1.3.4 ION CURRENT DETECTORS

Once ions have been separated according to their m/z values they must be detected (block 4 in Figure 1.2), and the resulting signals amplified and stored. Although there are multiple forms of detectors, the principle of operation is essentially the same. In *electron multipliers*, ions arriving at the point of collection hit a surface from which electrons are ejected. These electrons are then accelerated against a second surface from which an increased number of electrons are released. This multiplicative process is repeated numerous times, providing an increasing cascade of electrons, resulting in a current that is, in turn, amplified electronically and then recorded by the data system.

Ion detection is different for the FT instruments. Here ions move past a surface and in doing so induce an alternating (sinusoidal) electrical current the frequency of which is proportional to the m/z of the ions, while the intensity of the signal reflects the concentration of the analyte. The resulting *image current* is a composite of all the different induced current frequencies, and must be deconvoluted mathematically, using Fourier transform methodology, to obtain a mass spectrum.

TABLE 1.4
Comparison of Frequently Used Mass Analyzers

Parameter	Quadrupole (Q)	Time-of-Flight(TOF)	Orbitrap	FT-ICR
Ion beam	Continuous	Pulsed	Pulsed	Pulsed
Separation mechanism	Scanning	Time-of-flight	Oscillation	Rotation
Detection	Ion current	Ion current	Image current	Image current
Resolution	Unit	$>10^4$	$>10^5$	$>10^6$
Analyzer pressure (Torr)	10^{-5}	10^{-7}	10^{-10}	10^{-10}
Mass accuracy (<1 kDa)	1 Da	2 mDa	0.5 mDa	0.1 mDa
Mass limit, kDa ($z = 1$)	4	>350	4	4
Rate of spectra collection (Hz)	Scan 20 (often 1)	10,000 (often 1)	10	<1
Benefits	Scanning and selected ion monitoring	Fastest high-resolution spectra collection; highest mass range	High resolution	Highest resolution
Limitations	Nominal mass only	No selected ion monitoring	Resolution decreases with data collection speed and increasing mass	Resolution decreases with data collection speed and increasing mass; superconducting magnet
Current (2013) cost (US$)	$50,000+	$200,000+	$250,000+	$1,000,000+

1.3.5 VACUUM SYSTEMS

Mass spectrometers operate at high vacuum to prevent the loss of ions through collisions with neutral gas molecules or other ions. Vacuum is established by a combination of backing (roughing) and high-vacuum (usually turbomolecular) pumps. Most *backing* pumps are oil-filled *rotary vane pumps*, but the oil-free *scroll pumps* are now being used frequently. The latter have the advantage of reducing the number of background ions. The rotary or scroll backing pumps are used to bring the chamber to an initial vacuum at which point the turbomolecular pump(s) can begin to function. The second, equally important, function of the backing pumps is to remove the exhaust from the turbomolecular pumps during normal instrument operation.

Turbomolecular pumps are, in essence, sophisticated high-speed fans that draw air molecules out of the various chambers of the instrument. These pumps are convenient because they cycle rapidly and, unlike the older *diffusion* pumps, do not contain oil that can contaminate the vacuum chambers. Along with the pumps, the vacuum systems include various interconnected vacuum chambers, valves, and pressure gauges.

Differential pumping is the term used to describe the process by which different pressures are maintained in various locations within the instrument, e.g., during the desolvation of ESI droplets and in the collision-induced dissociation cells used in MS/MS instruments. As dictated by the requirements for differential pumping, vacuum systems are designed as a series of interconnecting chambers with low conductance connections, between which ions can be moved efficiently, while the pressures necessary for the operation of the individual components are preserved. Typically, pressures of ~10^{-5} to 10^{-7} Torr (~10^{-10} Torr in FT instruments) are required in mass analyzers.

1.3.6 DATA SYSTEMS

Computers are integral components of all modern mass spectrometer systems where they have two major functions: instrument control (including data acquisition) and data handling. Dedicated computers control most instrument operation functions, ranging from relatively simple parameters, such as tuning and calibration, to more complex decision-making processes required during data acquisition.

The combination of advances in instrument design and electronic hardware has increased dramatically the rates of data production and acquisition, particularly for TOF and FT systems. It is a relatively simple matter to collect full spectra with unit resolution in 0.5 s, as is the case for quadrupole instruments, compared with the collection of data on a TOF system where 10,000 data sets can be generated in 0.5 s with resolutions up to 60,000. In addition, the size of data files has increased rapidly, e.g., long LC-MS/MS runs can be on the gigabyte scale.

The second stage of computing is data handling, and here again, different levels of sophistication are dictated by the information required. The same computer that is controlling the instrument can often acquire data, produce spectra, and provide both interpretation and quantification. The amount of data accumulated in proteomics, and the complexities associated with the processing of such data, have led instrument manufacturers and independent software developers to produce highly sophisticated packages for data handling and interpretation (and display) that require the transfer of data sets onto a second computer. Web-based libraries and advanced software tools are increasingly becoming essential for efficient data interpretation, e.g., for protein identification based on the amino acid sequences of peptides derived from MS/MS experiments.

1.4 DEFINITIONS CONCERNING INSTRUMENT PERFORMANCE

What is resolution? The *resolution* of one mass from another and the sensitivity of ion detection are arguably the two most important performance parameters of a mass spectrometer. Resolution is a measure of the ability of a mass analyzer to separate ions with different *m/z* values. Resolution is determined experimentally from the measured *width* of a single peak at a defined percentage height of that peak and then calculated as *m*/Δ*m*, where *m* equals mass and Δ*m* is the width of the peak. The full width of the peak at half its maximum height (FWHM) is the definition of resolution used most commonly (Figure 1.8). For example, when an ion of mass 600 has

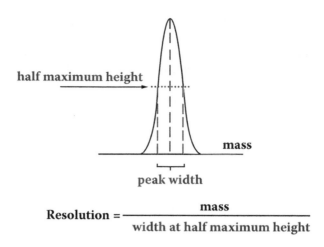

$$\text{Resolution} = \frac{\text{mass}}{\text{width at half maximum height}}$$

FIGURE 1.8   Calculation of resolution (full width at half maximum (FWHM) definition).

a peak width (Δm) of 0.075 at half its maximum height, the resolution is 600/0.075, or 8,000. There are no exact definitions of where low resolution (<2,000) or high resolution (>5,000) starts with respect to either the data generated or the types of mass analyzers.

What is resolving power? While resolution is usually a large number, *resolving power* is the inverse of resolution, $\Delta m/m$, and thus a small number, often expressed in terms of parts per million (ppm). Accordingly, the resolving power required to separate two compounds is the difference between the masses of the two ions divided by the mass of one of the ions. The product is multiplied by 10^6 so the value can be expressed in parts per million (ppm). For example, separating the molecular ion of cholesterol, MW 386.3845 ($C_{27}H_{46}O$), from that of ergostane (a hydrocarbon), MW 386.3913 ($C_{28}H_{50}$), a resolving power of (0.0364/386.3549) × 10^6 or 94 ppm is required.

It is important to understand how resolution and resolving power are used in reporting mass spectrometric data. The terms *resolution* and *resolving power* are defined in some published lists as they are above, while in others the definitions are reversed. On occasion, the two terms are used interchangeably. Having multiple definitions can be confusing.

What is accuracy of mass measurement? The difference between the calculated and experimentally determined masses of an ion provides a numerical measure of the *accuracy of the experimental data*, expressed in terms of millidaltons (0.001 Da, mDa), millimass units (mmu) (where mDa ≡ mmu), or parts per million (ppm). When the objective is to determine the elemental composition of an analyte, it is important that the mass measurements should be highly accurate, within a few mDa. While such measurements are usually associated with data acquired at high resolution, it is noted that relatively high mass measurement accuracy may also be achieved at low resolution, as long as care is taken to avoid interferences from closely adjacent peaks.

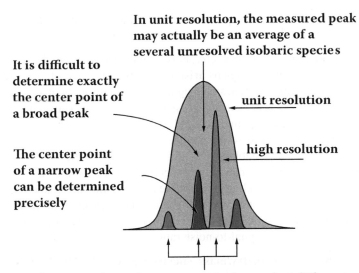

In unit resolution, the measured peak may actually be an average of a several unresolved isobaric species

It is difficult to determine exactly the center point of a broad peak

unit resolution

The center point of a narrow peak can be determined precisely

high resolution

Isobaric ions have the same nominal mass but different empirical formulae, and thus different accurate masses. Isobaric peaks can be resolved by high resolution.

FIGURE 1.9   High resolution improves the process of accurate mass measurement.

Why does high resolution improve the accuracy of exact mass measurement? There are two reasons (Figure 1.9). First, increased resolution permits the separation of isobaric ions. This is important because even small amounts of isobaric background ions will distort the measured mass of the analyte as the masses of the interfering ions are averaged into the measured mass. Second, the broader a peak is, the more difficult it is to precisely determine its center point (center of gravity).

What is upper mass limit? This is the *maximum measurable m/z* that can be observed in a given mass spectrometer. This is obviously a critical consideration in the selection of an instrument for a particular project. The mass range of an instrument is not directly related to its resolution, so poor resolution at high mass can severely reduce the useful portion of the mass range.

What is scan speed? Scan speed is the *rate* at which mass spectra are acquired, and is measured in masses/unit time. This term is usually applied to those analyzers where certain instrumental parameters, such as a set of voltages that control an electrical field, must be swept (scanned) continuously to transmit ions with successively higher, or lower, *m/z* values onto a detector where their intensities are recorded as a function of time. Quadrupoles and quadrupole ion traps, where voltages are scanned, and magnetic sector systems, where the magnetic field is varied, are scanning instruments. In contrast, nonscanning analyzers record data for all masses, at essentially the same time, without scanning any instrument parameter, e.g., time-of-flight instruments and ion traps that use Fourier transform data analysis. Although the nonscanning instruments do not have a scan rate as such, the time period over which data are collected (often called a scan) has important implications for instrument operation.

What is noise? In mass spectrometry, noise can be classified as *chemical* or *electronic*. The former consists of signals arising from the presence of chemicals other than the analyte in a sample. Often small signals are seen throughout a spectrum, and these are usually defined as noise, whereas specific interferences to an analyte signal are defined as interfering compounds; e.g., MADLI matrix ions can supply both types of chemical interference, as there are many small signals throughout the spectrum, but also specific ions attributable to the molecular mass of the matrix. Electronic noise is an inevitable consequence of the functioning of electronic circuitry and the result of multiple factors, including temperature and manufacturing tolerances.

Whatever the source, or combination of sources, noise places a limit on the intensity of an analyte signal that can be observed and recorded. The ratio of analyte to noise intensity is called the signal-to-noise ratio.

What is mass spectrometry/mass spectrometry (MS/MS)? MS/MS, often referred to as *tandem* mass spectrometry, is an important technique that acquired its name because of its ability to conduct two mass separations consecutively within the same instrument. MS/MS is based on analyzing the products of *controlled collisions* between selected ions (*precursors*) and neutral gas molecules in pressurized *collision cells* that are placed in specific regions of the instrument so that the collisions occur between two mass analyzers.

In both *tandem-in-space* and *tandem-in-time* instruments, the most common experiment is for the first analyzer to select specific ions from the total ion beam arriving from the ion source. Next, the selected ions undergo *collision-induced dissociation* (CID) in a pressurized cell followed by the analysis of the *product* ions in the second analyzer. In tandem-in-time the same analyzer is used for both scans, but at different times. The resulting product ion spectra (or *precursor* and *neutral loss* spectra in other forms of MS/MS analysis) provide vital structural information for the identification of small molecules (such as drug metabolites) as well as complex biomolecules. *Selected reaction monitoring* (SRM), another mode of MS/MS operation, provides highly specific and sensitive quantification of target analytes.

1.5   DEFINITIONS CONCERNING APPLICATIONS

What is meant by the diversity of mass spectrometry? Mass spectrometry does not deal with a well-defined property of molecules (such as infrared or ultraviolet absorption), and thus there is *no such thing as a single correct mass spectrum of a compound*. Instead, the diversity of mass spectra means that significantly different types of mass spectra may be obtained for an analyte, depending on the technique employed. This is a major advantage because judicious selection of the ionization method, type of analyzer, and experimental conditions greatly expand the applicability of the technique, enabling the acquisition of information concerning the identity, structure, and quantity of analytes. There are numerous applications where this diversity is advantageous. For instance, when small molecules are studied, different ionization/fragmentation methods can be used in combination to obtain both molecular mass and structural information. Another example is the analysis of peptides derived from proteins where the choice of fragmentation method can be used to determine the sequence of the amino acids as well as the presence and location of post-translational modifications. In addition, molecular masses of intact proteins can often be measured within a few Da of the calculated value. No other analytical method can determine mass to this level of accuracy.

The ability to interface mass spectrometry with a separation method, such as GC, LC, and CE, along with advances in automation, enables another level of diversification whereby a second dimension of component separation is added, thereby increasing the types of analytical questions that can be addressed for complex mixtures.

What are analytical specificity (selectivity) and sensitivity? The meanings of *specificity*, *selectivity*, and *sensitivity* vary significantly depending on the areas to which they are applied, e.g., in analytical methodology, clinical diagnosis, environmental contamination, etc. (Table 1.5). *Specificity* and *sensitivity* are terms related to instrument performance. However, the typical user of mass spectrometric data is usually more interested in *method-related* definitions of specificity and sensitivity that determine the usefulness of a particular technique with respect to the problem being investigated. For example, what is the minimum concentration that can be observed for a drug and its metabolites in a tissue, or for an environmental contaminant in a

TABLE 1.5
Specificity, Sensitivity, and Limit of Detection

Specificity	Description
Mass spectrometer	Ability to obtain unique identifiers for an analyte, e.g., *m/z* values of the molecular species or characteristic fragments.
Analytical method	Ability to observe analytes without detectable interference from the sample matrix.
Clinical	Fraction of false positive responses, e.g., misdiagnosis of a disease state when none exists. This fraction should be a very small number, i.e., a low failure rate.

Sensitivity	
Mass spectrometer	Instrument response to the detection of an ion. May vary according to the *m/z* value.
Analytical method	Ability to detect an analyte in a matrix. The *incremental sensitivity*, i.e., the minimum change in a concentration that can be reproducibly observed, is often an important consideration.
Clinical	Fraction of false negative responses, e.g., number of disease states that are missed, usually reported as the percentage of positive results or a positive fraction, e.g., 99% or 121/130.

Limit of Detection	
Mass spectrometer	Least amount of analyte that can be observed above the instrument noise (often a 3:1 ratio is used).
Analytical method	Least amount of analyte that can be observed after it has been recovered from a matrix, e.g., from serum or an environmental sample.

lake sediment? A different set of definitions of specificity and sensitivity is applicable in clinical applications and concerns the number of false positive/false negative results, e.g., for a putative disease marker.

Specificity and selectivity are used interchangeably, including in most journals. Despite common use, the meaning of *specificity* is controversial, particularly when contrasted to *selectivity*. The IUPAC recommends (provisionally) the use of *selectivity* and discourages using *specificity*, suggesting that while a method is either specific or not, selectivity is quantifiable.

From an analytical point of view, selectivity refers to the extent to which a given method can determine the presence of a particular analyte in a matrix or in a mixture without interference from other components. For example, the strategy of SRM provides significantly higher selectivity than techniques based on chromatographic retention times only.

Selectivity is of relevance in the statistical evaluation of mass spectrometric data when scoring each comparison between experimental data and a model. Selectivity is the ratio of the number of false results not rejected to the total number of false results and should be a small number. This number is used to determine significance, which is 1 − selectivity (preferably close to 1). In this respect, selectivity should be compared with sensitivity, which is the ratio of the total number of true results not rejected to the total number of true results (preferably a large fraction/percentage).

What are limit of detection (LOD), limit of quantification (LOQ), and incremental sensitivity? The *limit of detection* (LOD) is the smallest quantity of an analyte that can be detected, with respect to either the background noise of the instrument (*instrumental LOD*) or the matrix (biological or environmental) in which an analyte is to be determined (*analytical LOD*). A signal-to-noise ratio of 3 is frequently used as the definition of LOD. The *limit of quantification* (LOQ) of a given analytical technique is usually taken to be about three times its LOD. *Incremental sensitivity*, which is of major practical importance, is defined as the minimum difference between two concentrations that a particular technique can distinguish reliably (Table 1.5).

Experimental conditions, including the type of instrumentation used and the matrix in which an analyte is assayed, significantly affect the sensitivity/specificity/LOD/LOQ of a measurement. Reported data should therefore include sufficient information to assist in both the interpretation of the results and, if necessary, in repeating or carrying out similar experiments.

1.6 INFORMATION FROM MASS SPECTRA

Mass spectrometers answer the basic questions of *what* and *how much* is present of a given analyte by determining ionic masses and their intensities. Accordingly, the major areas of applications of mass spectrometry are in both qualitative (including structural) analysis and quantification.

The choice of an ionization method, and even how it is used (*experimental conditions*), can have dramatic affects on the spectra obtained. For example, EI spectra are highly reproducible and contain structural information but may not give the molecular masses of the analytes. In contrast, CI spectra provide molecular mass information,

[M + H]+ ions, and sometimes structural data, depending on the reagent gas used. ESI spectra are usually comprised of only ions associated with the molecular mass of the analyte and provide no structural information, although some fragmentation can be induced by changing the voltage at the point where ions enter the vacuum chamber (*cone voltage*).

The judicious selection of an ionization method is a very important step toward a successful analysis. Indeed, mass spectrometric results should always be described with reference to the ion source employed. The choice of an ionization method may also depend on other factors, such as the molecular mass and polarity of the compound and whether GC or LC is being used for the initial separation of the constituents of mixtures. It is emphasized that mass spectra are usually reproducible when experimental parameters are duplicated. The consequence of the wide variety of choices of techniques and strategies available is that in mass spectrometry, compared with other spectroscopic techniques such as UV and infrared (IR), it is particularly important to be familiar with alternative instrument types, analytical techniques, and operational parameters.

Low-resolution analyzers, notably quadrupoles and the related ion traps, provide unit mass resolution, separating masses that differ by 1 Da. These instruments, particularly when coupled with GC or LC, are widely used for the quantification of target analytes. The initial chromatographic step serves to separate isobaric (same nominal mass) and other constituents from the analyte. Another major area of application for which unit resolution spectra are adequate is in the rapid confirmation or identification of analytes using the large libraries of EI spectra.

The *structural elucidation* of low molecular mass analytes is based on the interpretation of mass spectral fragmentation patterns obtained by EI. Decades of experience have resulted in numerous books that provide data and strategies for the interpretation of EI spectra for a wide variety of organic compound classes, ranging from simple hydrocarbons to complex alkaloids. Further structural information may be obtained from the application of MS/MS techniques, where selected ions from the primary fragmentation of the analyte in the ion source undergo secondary or higher-order fragmentations using CID. This type of fragmentation follows different rules from EI and provides an alternative set of data with distinct structural information. Still, it is rarely possible to determine the structure of complex organic or biological materials by MS alone, without complementary information from nuclear magnetic resonance (NMR), UV, and IR spectroscopy as well as other analytical methods.

High-resolution analyzers separate *isobaric ions*, thereby significantly increasing selectivity. For example, the resolution needed to separate the isobaric ions of hexachlorobiphenyl ($C_{12}H_4{}^{35}Cl_6$) at *m/z* 357.8444 and a specific isotope of pentachlorodibenzo-*p*-dioxin ($C_{12}H_3{}^{35}Cl_3{}^{37}Cl_2O_2$) at *m/z* 357.8519 is $m/\Delta m = 357.8519/0.0075$ or ~48,000, a difficult but doable task. It should be emphasized that isomers, by definition, have the same empirical formulae, and therefore chromatographic separation is required to observe and analyze the components of such mixtures.

The measurement of the masses of ions with accuracies to three or four digits beyond the decimal point allows the determination of possible elemental compositions of the ions. The basis of such determinations is that elemental masses (except carbon) and, consequently, the ionic masses of analytes, do not have integer values. An important application of high-resolution MS, with accurate mass measurement, has been the identification of unknown analytes based on the determination of their elemental compositions. The required accuracy increases rapidly as the mass increases; thus this approach is usually restricted to analytes of <1 kDa. However, because the types of components of which biopolymers are composed are limited, e.g., amino acid residues in proteins, and because of technological advances in TOF-MS and FT-MS instrumentation, the mass range over which accurate mass measurement can be made in specific applications, such as in peptide sequencing, has been extended (Section 3.1.3).

The *quantification* of selected constituents present in trace quantities (e.g., a drug in plasma or an environmental contaminant in a river sediment) requires an ability to select specific ions with quadrupole analyzers in either MS or MS/MS mode. When single analyzer instruments are used for this purpose, the technique is termed *selected ion monitoring* (SIM) and the analyzer serves as an ion selection device that allows only one (or a few) preselected ions of specific *m/z* value(s) to reach the detector, excluding all other ions present. When MS/MS is used to monitor specific ions with triple quadrupoles (QqQ), the strategy is called *selected reaction monitoring* (SRM). In SRM a precursor ion, selected in the first quadrupole (Q1), is allowed to undergo CID in the second, rf-only region (often a second quadrupole, q2), with only the product ion of a characteristic fragmentation being transmitted by the third quadrupole (Q3) to the detector (Section 3.3.3.1).

The selectivity provided by SIM and especially SRM is applicable to the quantification of a wide variety of compounds, in almost any matrix, including body fluids and tissues, with minimal sample preparation. The value of SIM and SRM in quantification is that the analyzer is dedicated to acquiring data on only the ions of interest, with all other ions excluded, resulting in significant improvements in detection limits because of the increase in the amount of time that an ion is observed and improvement of the signal-to-noise ratio.

Table 1.6 provides examples of some objectives of mass spectrometry for both qualitative and quantitative applications.

TABLE 1.6
Some Analytical Objectives Using Mass Spectrometry

Application	Process
Qualitative	
Known compounds (confirmation of identity)	Compare with standards
	Compare against libraries
	Small molecule (<1 kDa)
	Biopolymer databases
	Compound interaction (e.g., protein:drug complex)
Unknown compounds (determination of identity)	Molecular weight
	Accurate mass for determining empirical formula (<1 kDa)
	Structural information from fragmentation pattern and isotopic composition
	De novo sequencing (proteins)
Quantitative	
Known compounds, or estimations of unknowns of potential importance	Small molecules (e.g., environmental pollutants, pharmaceuticals)
	Proteins (e.g., differential expression)
	Mixture composition

1.7 DIVERSITY AND SCOPE OF APPLICATIONS

Although a given analytical protocol can be carried out on various instrument configurations, it is relevant to consider which instrument type is best suited to solve the particular problem. The nature of samples is tripartite, including the molecular mass and polarity of the analyte, and the complexity of the sample matrix. Perhaps the most straightforward approach is to consider samples according to their molecular masses, going from low to high. Coincident with molecular mass is the degree of polarity. Low molecular mass compounds may be polar or nonpolar, but most high molecular mass species are polar. The complexity of a sample can range from a single compound, e.g., a purified compound from a chemical reaction, to multiple species in complex matrices, e.g., proteins in biological tissues or trace contaminants in environmental samples.

Identification or quantification of molecules up to 1 kDa represents the majority of mass spectrometric analyses. The simplest analysis is probably that of a single product from a synthetic chemical reaction where the required information is only the molecular mass. Such samples used to be introduced using a direct insertion probe. Nonpolar compounds are now analyzed by GC-MS with EI or CI. For polar compounds the usual method of introduction is in a liquid flow, e.g., in methanol (without an LC column), that then is vaporized and the analyte ionized using ESI or APCI. This approach is termed *flow injection analysis* (FIA). While quadrupole analyzers are applicable for ether GC-MS or FIA, TOF and FT instruments provide the added dimension of accurate mass determination.

The availability of autosamplers, combined with GC-MS, LC-MS, or ESI with FIA, has removed the need for instrument operators who analyze one sample at a time. Instead, a large number of samples can be analyzed unattended, including during overnight operations. Instrumentation and software are now available to assist users who need data in a frequent and ongoing manner but do not have formal training in mass spectrometry. An example of this type of user is the synthetic chemist for whom nominal mass data are usually sufficient; e.g., if m/z 359 is expected and m/z 401 is observed, then it is clear that the wrong (unexpected) compound was synthesized. Nominal mass determination is obviously sufficient in this case, although accurate mass measurement (to obtain the empirical formula) can be important to understand the unexpected result. Accurate mass measurement is useful in other instances: consider that the ions $[M + K]^+$ and $[M + O + Na]^+$ will both be observed at the same nominal m/z, $[M + 39]^+$. However, these two ions will have different accurate masses (e.g., $[C_{10}H_{20}N_2 + O + Na]^+ = 197.0691$ Da, while $[C_{10}H_{20}N_2 + K]^+ = 197.0481$ Da); obtaining these data will make it apparent immediately whether M has or has not been oxidized to MO.

The next level of complexity is the analysis of compounds in complex matrices such as blood plasma, biological tissues, or environmental samples. Here the class of compounds of interest needs to be recovered from the matrix using, for instance, solvent or solid phase extraction. The resulting extracts are usually still complex mixtures that are best separated chromatographically prior to introduction into the mass spectrometer. The polarity of the analytes will determine whether GC or LC is the appropriate technique. The versatility of QqQ instruments is particularly suited to such analyses, with SRM being used for quantification, while progeny, precursor, and neutral loss scans (Section 3.3.3.1) can be used to search for classes of compounds related to the initial analyte, such as drug metabolites.

Chromatographic separation may not be adequate for some highly complex mixtures of molecules, e.g., crude oil, the by-products of the chlorination of water, or the humic and fulvic acids found in river water. As an example, the characterization of crude oil may be needed to determine the origin of a sample for provenance or enforcement purposes or to obtain information to assist in developing improved refining processes. Here the use of multiple ionization methods, combined with the ultimate resolving power of FT-ICRMS, can provide compositional information. This can include determining the sulfur content of the mixture, based on resolving and specifically searching for ions that contain the ^{34}S isotope using the mass defect resulting from the inclusion of this isotope and for classifying compounds with different levels of aromaticity or aliphatic side chain lengths by specific processing of the spectral data.

Common analytes among large molecules are the biopolymers: proteins, nucleic acids, and carbohydrate complexes. These compounds, particularly proteins, can be analyzed either as intact species or, after digestion, as peptides. Of importance in analyzing protein digests are LC-ESI-MS/MS and MALDI-TOF/TOF instruments where CID is used to provide amino acid sequence information (Sections 3.3.2.1 and 3.5.1.1). The locations of post-translational modifications (PTMs) of proteins, that are vital to the understanding of protein function, can be obtained using MS/MS techniques such as electron transfer dissociation (ETD) and electron capture dissociation (ECD) (Section 3.5.1.5). Among other biopolymers, nucleic acids are linear structures and can be sequenced by MS/MS similarly to proteins (Section 3.5.2). In contrast, most sugars are branched structures that are difficult to characterize and require multiple levels of MS/MS (MS^n) analysis (Section 3.5.3).

The top end of the mass scale includes *intact* proteins, e.g., the investigation of noncovalent interactions between a protein and small molecules, or between two

---

TABLE 1.7

Scope of Mass Spectrometry—Examples of Applications

Field	Process
Chemistry	Identification and quantification of components from reaction mixtures
	Process monitoring
Environmental	Identification and quantification of contaminants
	Distribution of contaminants (e.g., biomagnification)
Pharmacology	Drug and metabolite identification/quantification (pharmacokinetics, pharmacodynamics, pharmacogenetics)
Biology	Identification of proteins and other biopolymers
	Identification of mutation altered proteins
	Protein function control through post-translational modification
Medicine	Identification of disease markers
	Differential protein expression

or more proteins (Sections 3.5.1.7 and 4.12). Another example is the top-down sequencing of intact proteins using FT-ICR instruments (Section 3.5.1.3).

Table 1.7 provides various examples of the range of applications to which mass spectrometry can be applied.

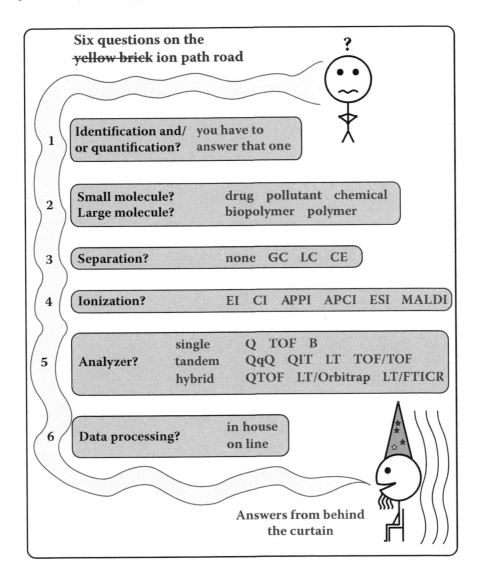

2 Instrumentation

This chapter revisits the instrumentation outlined in Chapter 1, but now taking a closer look at the multiple formats available for each component of a mass spectrometer. It is appropriate to show the overall layout of a mass spectrometer again (Figure 2.1), but with more detail, including different approaches to sample introduction, methods

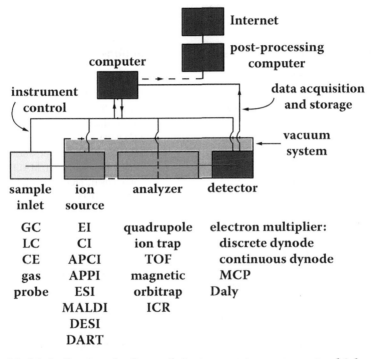

Multiple forms exist for each instrument component, which can usually be mixed and matched.

Analyzers can be used in single, e.g., Q or TOF, or in multi-analyzer formats, e.g., QTOF and TOF/TOF, with a collision cell incorporated between the two analyzers.

The computer controls the instrument, acquires data, and enables routine data processing, e.g., producing and quantifying spectra.

Complex data processing, e.g., protein identification, is often moved to a separate computer and uses on-line libraries.

FIGURE 2.1 Block diagram of a mass spectrometer (expanded).

of ionization, and types of analyzers. The functions of the components shown in the figure are summarized in Table 2.1.

An aside is needed to consider the molecular mass (weight) of analytes, as this has consequences for both the type of instrumentation used and the analytical technique selected. Molecules are often referred to as small or large, but the distinction between these categories is a matter for conjecture. It is generally accepted that analytes with masses of <1 kDa are *small* molecules. Small proteins with a mass of ~5 kDa, such as insulin, mark the lower boundary of the *large* molecule category. The intermediate range of 1–5 kDa is a gray area that depends on the objective of the project. For instance, when a molecule in the 1–5 kDa range is a synthetic product, it is more likely

TABLE 2.1
Mass Spectrometer Components and Their Functions

Component	Type	Function
Sample introduction	Gas (batch) inlet	Entry of a regulated sample stream,
	Direct gas introduction	e.g., calibration compound
	API	GC interface
		LC interface
Ion source	EI	Generation of ions from samples either
	CI	under vacuum (EI, CI, MALDI) or at
	APCI	atmospheric pressure (ESI, APCI, APPI,
	APPI	DESI, DART)
	ESI	
	MALDI	
	DESI	
	DART	
Analyzer	Quadrupole	Separate ions based on their *m/z* ratios
	Ion trap (quadrupole-based)	
	time-of-flight	
	Ion trap (Fourier transform based,	
	orbitrap, or ICR)	
Collision cell	Gas cell	Provide fragmentation for MS/MS
(Ion) detector	Electron multiplier:	Record the arrival and intensity of ions
	Discrete dynode	emerging from the analyzer
	Continuous dynode	
	MCP	
	Daly	
Vacuum system	Rotary pump (oil based)	Provide initial vacuum and backing for
	Scroll pump	turbomolecular pumps
	Turbomolecular pump	Create high vacuum, 10^{-5} to 10^{-10} Torr to
		allow movement of ions without loss
		through collision
Computer	PC	Control instrument operation
(acquisition)		Acquire and process data
Computer	PC	Data processing and library comparison
(post-processing)	Internet/www	

to be considered a large molecule. When such a molecule is a peptide derived from a protein for sequencing, then the analysis is more akin to that of a small molecule, presumably because structural information is the objective. There is also a historical element in the distinction between the size classes of molecules, as masses above 1 kDa were mostly inaccessible to mass spectrometry until electrospray ionization (ESI) and matrix-assisted laser desorption/ionization (MALDI) were developed.

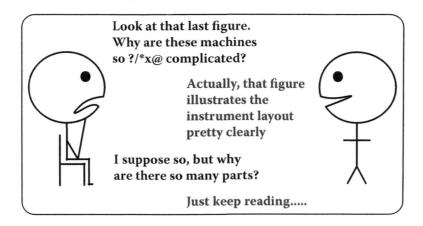

2.1 SAMPLE INTRODUCTION

The main function of the sample introduction system is to introduce an appropriate quantity of a sample into the ion source so that its composition represents that of the original sample as accurately as possible. The range of samples that can be analyzed by mass spectrometry is immense, varying from simple gases to large protein complexes. Samples may consist of a single species or highly complex mixtures. Although pure samples or simple mixtures of low molecular mass analytes can often be admitted directly into an ion source, mass spectrometers have, in general, a limited ability to characterize complex mixtures. Therefore, the components of mixtures must be separated (usually by chromatography) prior to their ionization and analysis.

2.1.1 PURE SAMPLES OR SIMPLE MIXTURES

Gases, products from chemical reactions, and compounds used to calibrate mass scales are usually introduced into the ion source directly.

2.1.1.1 Introduction of Gases and Volatile Liquids Using Batch Inlets

Batch inlets should enable a constant and reliable flow rate of the sample during the analysis, and be inert to the analytes. Gases and volatile liquids are introduced into EI or CI sources from a glass or metal reservoir of known volume (that can be heated to 350°C) by way of a gold or steel foil that has one (or several) pinhole(s) (0.005 to 0.02 mm diameter). The introduction of liquids can be controlled by placing the sample at the bottom of a small tube (immersed into dry ice or liquid nitrogen) connected to a pumping manifold. Sample utilization in batch systems is poor (~0.01%) because an excess of sample is required to maintain the necessary vapor pressure so that there is a constant flow of the analyte into the ion source. Still, because sample flow can be controlled accurately, the method is appropriate for routine quantitative analysis of organic and inorganic gases, volatile liquids, and with appropriate heating of the reservoir, solids with limited volatility, such as waxes and tars. These types of measurements are well suited to industrial process control applications.

The most common use of batch inlets is to introduce a controlled flow of compounds for calibrating the mass scale. Frequently used calibration compounds include perfluorotributylamine (FC-43, heptacosa) and perfluorokerosene (PFK); both are effective with electron ionization (EI) but give limited responses in chemical ionization (CI). If used for calibration in CI, or to provide lock masses (Section 3.1.1), the concentration of the reagent gas must be reduced. This reducation somewhat compromises the effectiveness of the CI process.

2.1.1.2   Semipermeable Polymer Membrane Inlets

Some polymers, e.g., silicone membranes, have permeabilities that are significantly different for volatile organic compounds relative to water or inorganic gases, thus permitting the selective introduction into the ion source of certain analytes present in aqueos or gaseous streams. As the stream passes over the (heated) membrane the analytes are selectively absorbed into one face of the membrane and then desorbed from the other side by the vacuum inside the mass spectrometer. There is no need for sample preparation (including extraction or preconcentration), and analytes can be monitored and quantified in the parts per hundred to parts per trillion range. The avoidance of sample preparation has led to the use of *membrane inlet mass spectrometry* (MIMS) in an increasing number of divergent applications, ranging from on-site, on-line monitoring of industrial processes and waste effluents to physiological and metabolic studies.

2.1.1.3   Direct (Static) Insertion Probes

Direct insertion probes (also called *direct inlet probes*) use a small, one-end-closed, quartz, glass, or inert metal (platinum or gold) capillary into which a pulverized solid sample or one dissolved in a volatile solvent is placed. The probe is inserted into the instrument through a *vacuum lock* and placed close to the ion source. Samples are evaporated into the ionization region by heating the capillary up to 400°C. Spectra are obtained when adequate sample vapor pressure is reached at temperatures below those where the analytes decompose. The high vacuum in the instrument assists the vaporization; while controlling the rate of increase of the probe temperature enables microdistillation of crude mixtures. Sample utilization is good because analytes are volatilized only a few millimeters from the ionizing electrons (Section 2.2.1), or from the reagent gas ions in CI (Section 2.2.2.1). As little as 5 ng of analyte may yield usable information. A disadvantage of direct probes is that the mechanical vacuum lock through which the probe must pass is operated manually. Such a lock is liable to failure or to accidental vacuum loss during operation. The use of solids probes has diminished (almost to nonexistence) with the development of the atmospheric pressure ionization methods.

Flow injection analysis (FIA), using ESI or APCI (Sections 2.2.2.2 and 2.2.2.3), where samples are introduced in a liquid flow without a chromatographic column (and without vacuum locks), has become the modern equivalent of the direct insertion probe. In addition, FIA combined with API methods usually eliminates problems of sample decomposition and limited mass range encountered with solids probes.

2.1.2   CHROMATOGRAPHIC SEPARATION OF COMPLEX MIXTURES

The enormous synergistic analytical potential of combining the separative powers of gas chromatography (GC) and liquid chromatography (LC) with the identification and quantification capabilities of MS was realized decades ago. The main problem of combining the techniques is that both GC and LC operate at atmospheric pressure utilizing a mobile phase in which the analytes are carried, while MS is a vacuum-based technique. Although a detailed discussion of chromatography is beyond the scope of this book, a brief description of the technique is warranted because of its analytical importance and the consequences for interfacing the two instruments.

In chromatography, complex samples are introduced onto a column comprised of a solid support that is coated with a stationary phase (having a special chemical formulation) through which a *mobile phase* (a gas in GC and a liquid in LC) of variable composition passes. The premise is that the components of the injected mixture are adsorbed initially onto the stationary phase followed by a sequential release of analytes into the mobile phase as its properties are altered by temperature (in GC) or solvent composition (in LC).

Because the vast majority of samples are complex mixtures, they generally require the separation of their components, by GC or LC, prior to their introduction into the ion source. GC is usually carried out on fused silica capillary columns. LC is available in two formats: in conventional LC the flow rates are 0.1–1.0 ml/min, while *nano*-LC operates at sub µl/min flow rates. Capillary electrophoresis (CE) can be interfaced to mass spectrometers (similarly to LC). Thin-layer chromatography (TLC) is compatible with the newer surface ionization methods.

GC-MS and LC-MS systems have been described variously as combinations in which either the MS serves as a sophisticated detector for LC or the chromatograph is an inlet for the mass spectrometer. While this may appear a semantic distinction, it is more a reflection of how the mass spectrometer is being used as well as the background of the individual using the instrument. For example, the mass spectrometer may be considered a detector when it is used to quantify the amount of a drug present in plasma samples. In contrast, the chromatograph serves as the sample introduction technique when the objective is to identify and determine the structure of a drug metabolite. Of course, mass spectrometrists are likely to consider the mass spectrometer as the more important component of the combined instruments, while chromatographers will have the opposite view. In reality, each is vital to the other.

2.1.2.1   Gas Chromatography–Mass Spectrometry (GC-MS)

This technique was originally referred to as *gas liquid chromatography* (GLC); however, the name has been shortened to *gas chromatography* (GC). In GC the process of separation is based on a two-step sequence. First, the components of a mixture are adsorbed onto a coating that lines the inner wall of a column (usually a long fused silica capillary) located in an oven. Next, the oven is progressively heated and the adsorbed components are sequentially transferred into a gaseous, inert, mobile phase (usually helium) (Figure 2.2). The coating, which is chemically bonded to the inner wall of the column, is called the stationary (liquid) phase. The most common stationary phase, which is suitable for the majority of analytes, is a dimethyl-siloxane polymer where 5% of the methyl groups are substituted with phenyl groups. The use of other, specialized, phases depends on the nature of the

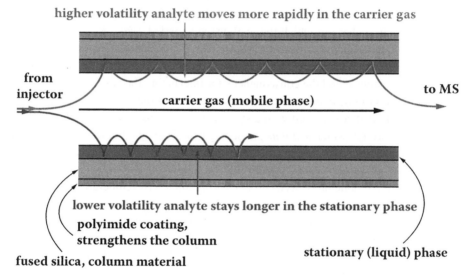

Analytes condense at the entrance of the column and are
   subsequently separated based on their molecular mass
   and polarity.
These properties determine analyte volatility and, as a
   result, the residence times in the stationary (liquid) phase
   (usually an organosilicon oil) and the gaseous mobile
   phase (usually He).
More volatile components elute first as they are
   carried through the column by the carrier gas at lower
   temperatures.
Increasing the oven temperature enables the transfer of
   compounds with higher boiling points from the stationary
   phase into the vapor phase and their elution from the column.

FIGURE 2.2   Separation process on a fused silica capillary GC column.

analytes; e.g., alumina plot columns are used for gases and organic compounds with low molecular masses (<100 Da).

The temperature at which a compound elutes from the column is a function of the vapor pressure of the analyte, which is derived from a combination of its molecular mass and polarity. The higher the vapor pressure of an analyte, the more rapidly it will traverse the column, carried by the mobile phase. As the column is heated progressively (*ramped*), the less volatile components of a mixture are eluted in sequence.

The commonly used fused silica capillary columns are highly efficient, enabling the rapid separation of dozens of compounds. Samples (1–3 µl), dissolved in a volatile solvent, are injected with a syringe into an injector volume heated to 150–280 °C. The volatilized components of the sample move out of the injector and are condensed at the beginning of the column that is held at a temperature (e.g., at 35 or 50 °C) much lower than that of the injector. Condensing the analytes focuses them at the beginning of the column and improves the efficiency of the separation. A temperature ramp, e.g., 50 to 290 °C at 10 °C/min, is then used to elute the analytes progressively from the column into a detection system, such as a mass spectrometer, flame ionization detector, or electron capture detector. The sample capacity of capillary columns is limited (pico- to nanograms per sample component), but this is partially compensated for because the columns produce peaks that are well separated and only a few seconds wide. The narrowing of the peaks results in an increase in the number of analyte molecules arriving in the ion source per unit time (i.e., at higher concentration); thus, there is an improvement of the signal-to-noise ratio and, consequently, the detection limit (Figure 2.3).

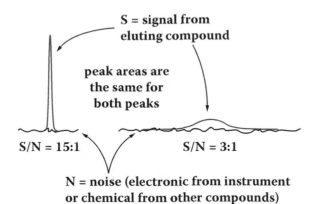

S = signal from eluting compound

peak areas are the same for both peaks

S/N = 15:1   S/N = 3:1

N = noise (electronic from instrument or chemical from other compounds)

The signal-to-noise (S/N) ratio improves when the width of the chromatographic peak is reduced.
The amount of material injected is the same in both cases shown. However, the number of ions arriving per unit time at the detector, i.e., the concentration, increases as the peak narrows. The higher concentration improves the S/N ratio.
In the illustration the detection limit is increased by a factor of five.

FIGURE 2.3   Signal-to-noise ratio vs. peak width.

A gas chromatograph and a mass spectrometer can be combined without major modifications to either instrument because the volume and flow rate of the gas used in GC (at atmospheric pressure) and the vacuum requirement of the MS ($\sim 10^{-7}$ Torr) can be reconciled. The interfacing was further simplified in the 1980s when GC transitioned from packed to capillary columns. Capillary columns were initially made from glass and were fragile. The subsequent development of fused silica capillary columns, which are structurally reinforced by a coating of polyimide, provided robust, flexible columns (Figure 2.2). Most capillary columns have internal diameters of 0.25 mm or 0.32 mm and use carrier gas flow rates of ~1 ml/min, as opposed to the 20 ml/min required for packed columns. The reduction in flow rates, together with improvements in the pumping designs of mass spectrometers, particularly differential pumping of the ion source (Section 2.5), permit the efficient removal of the carrier gas without compromising the vacuum necessary for the operation of the MS. Consequently, GC columns can be extended through a heated transfer region (held at ~250°C to prevent sample deposition) so that they abut the EI or CI source (Figure 2.4).

Although liquid injection is the usual method of sample introduction, there are other types of injector available. Techniques include *headspace analysis* for species with low molecular masses and high volatility, such as halocarbons in water; *thermal desorption* for analytes (e.g., pheromones) trapped by passage through a cartridge packed with a retentive material such as Tenax; and *solid phase microextraction* fiber elution for compounds that have been adsorbed onto the fiber from liquid or gaseous matrices.

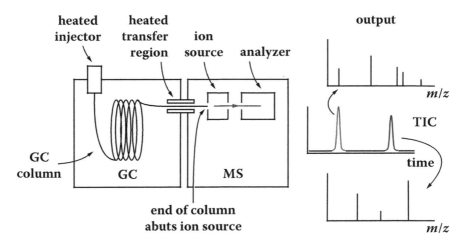

**Samples are introduced into the GC using a heated injector.
Components are separated on a column, according to a
    combination of molecular mass and polarity, and
        sequentially enter the MS source via a heated transfer region.
The analytical data consists of total ion chromatograms (TIC)
    and the mass spectra of the separated components.**

FIGURE 2.4   Gas chromatograph–mass spectrometer (GC-MS).

The need to volatilize samples has always been a limiting factor in GC-MS. As a rule of thumb, "one polar group good, two polar groups bad and don't run acids" is a useful guideline. Primary amines ($-NH_2$), hydroxyls ($-OH$), and carboxylic acids ($-COOH$) are the most relevant polar groups.

GC-MS remains a widely used analytical technique for the identification and quantification of a large variety of volatile or chemically derivatized compounds, including drugs, carcinogens, metabolic products, and environmental contaminants, usually with molecular masses of <600 Da (sometimes up to ~1 kDa). Methods of derivatization include the conversion of $-OH$ groups into trimethylsilyl ethers and the esterification of acids. Thousands of GC-MS applications have been reported.

2.1.2.2   Liquid Chromatography–Mass Spectrometry (LC-MS)

Most biological systems are aqueous and are composed of compounds that are involatile, polar, and often of high molecular mass; these characteristics prevent their analysis by GC-MS. Chromatography of such classes of compounds is possible with LC, thus the interfacing of LC with MS was essential to expand MS into biological applications. (LC is abbreviated from the original HPLC (high-performance/pressure liquid chromatography).)

The polarity of biological compounds was a challenge that led to the development of *reversed-phase* LC. The premise of this type of separation is that organic compounds in an aqueous mobile phase will adsorb preferentially onto an organic stationary phase coated on a solid support. Compounds are then eluted successively using a progressive transition (a *gradient*) of the composition of the mobile phase from aqueous to organic, e.g., from water to acetonitrile. The order of elution of the components of a sample is a function of the preference of particular compounds to be associated with either the stationary phase or the liquid (mobile) phase as the composition of the solvent changes during the gradient. Compounds elute in order of their polarity, with the most hydrophilic compounds eluting earliest (Figure 2.5). The stationary phase is nonpolar (in contrast to the polar phase used in *normal-phase* chromatography); this is why the method is known as reversed-phase. (The correct terminology is *reversed-phase* and not *reverse-phase*.) Among several available forms (chemistries) of stationary phase, the most common for analysis of molecules of <5 kDa is a C_{18} hydrocarbon chain. A C_4 or C_8 chain is often used for biopolymers. The porosity of the particles on which the stationary phase is coated is also varied, depending on molecular masses of the analytes. Molecules of <5 kDa are analyzed on particles in which the pore size is 100 Å, while a 300 Å pore size is used for biopolymers.

One problem with reversed-phase LC is that some very polar compounds, e.g., amino acids, dissolve preferentially in the mobile phase and do not adsorb onto the stationary phase. Such polar compounds are poorly retained, elute very close to the solvent front, and are separated ineffectively from each other or from inorganic salts that are similarly not retained.

In contrast to reversed-phase, the stationary phase in normal-phase chromatography is polar, usually silica or alumina, and uses nonpolar solvents, e.g., hexane and ethylacetate, that are not compatible with the API processes used in LC-MS. In normal-phase chromatography compounds elute progressively from the least to the most polar. The technique is not applicable to the highly polar compounds encountered

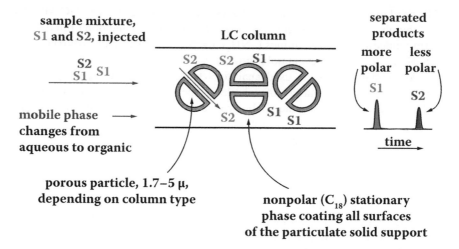

FIGURE 2.5   Separation process in reversed-phase liquid chromatography.

in biological systems, as these compounds frequently fail to elute and remain on the column. The development of hydrophilic liquid interaction chromatography (HILIC) chemistry has provided a form of normal-phase separation that utilizes aqueous liquid phases. This type of column chemistry is of interest because it can be used to analyze the very polar compounds that are not retained on C_{18} columns.

A distinction is also made in the categorization of liquid chromatography depending on the sizes of the particles that comprise the column packing material. LC uses 2.6–5 μ particle sizes with solvent pressures up to 3.1×10^5 Torr (400 bar, 6×10^3 psi). Separation efficiencies improve as particle size is reduced and are maintained over a wider range of flow rates, e.g., 0.2–1 ml/min. A consequence of using smaller particle sizes (1.7–2 μ) is that pressures up to 9.5×10^5 Torr (1,240 bar, 18×10^3 psi) are necessary to move the solvent through the column. This has led to a variety of potentially confusing names for systems that use <2 μ size particles and the necessary higher pressures, e.g., ultra-high-performance liquid chromatography (UHPLC) and ultra-performance liquid chromatography (UPLC).

Interfacing LC with MS is a much more difficult proposition than that for GC with MS because a milliliter of most organic solvents produces ~1,000 ml (and water ~1,600 ml) of vapor on evaporation at standard temperature and pressure. Such large volumes of vapor are incompatible with the vacuum requirements of mass spectrometers. Earlier attempts to interface LC with MS, including direct liquid introduction, the moving-belt interface, particle beam, and thermospray, were discarded with the development of the atmospheric pressure ionization (API) methods, electrospray ionization (ESI) and atmospheric pressure chemical ionization (APCI), where the solvent removal and sample ionization take place outside the vacuum chamber. The API techniques have also eliminated (or at least reduced) the problem of having to vaporize highly polar compounds without destroying them. LC separations can now be readily coupled with MS detection without major trade-offs in the achievable separations. Some needed compromises include the adjustment of methodologies to utilize more volatile buffers that can be removed effectively during the evaporative process, e.g., ammonium acetate instead of potassium phosphate, and the avoidance of trifluoroacetic acid (compromises ion formation because of the strength of the ion pairing with the analyte).

2.1.2.3 Capillary Electrophoresis (CE)

Capillary electrophoresis may be considered a specific form of LC where the separation is based on the charge and size of the analytes and their movement in a liquid, instead of the interaction of the analyte with solid and liquid phases. CE is based on the migration of ions in solution through fused silica capillaries (<100 μm diameter) under the influence of an electric field of high potential (kV). The charged analyte species migrates with different speeds in buffered solutions, resulting in a plug-like component profile due to the electro-osmotic flow created by the electric field. In the case of proteins the conditions are mild (without organic solvents), allowing the analytes to remain in their native configurations, permitting the detection of subtle changes, including post-translational modifications (Section 3.5.1.5). The usual interface with a mass spectrometer is an ESI source. Advantages of CE, compared to GC and LC, include high-efficiency separation (10^5 theoretical plates), small sample size, and low solvent consumption. Difficulties arise from the very low flow rates used (~200 nl/min).

2.2 ION SOURCES AND METHODS OF IONIZATION

Ion sources have two main functions: to generate ions that represent the components of a sample and to transfer all ions into the mass analyzer.

There have been numerous ionization techniques developed over the years, and their relative importance has varied as new ones have been commercialized. In all ionization sources energy is added to the neutral molecules present, including those from the analyte and from other species, e.g., co-extracted components and residual background. Agents used to produce ions include electrons, ions, electric fields, and photons.

Ions can be formed within the vacuum chamber of the mass spectrometer or outside the instrument at atmospheric pressure. Examples of *in vacuo* ionization are *electron ionization* (EI), *chemical ionization* (CI), and *matrix-assisted laser desorption/ionization* (MALDI). Ionization techniques carried out outside the vacuum system are collectively termed *atmospheric pressure ionization* (API). The most important API methods are *electrospray ionization* (ESI) and *atmospheric pressure chemical ionization* (APCI). Among other API techniques are *atmospheric pressure photo-ionization*

(APPI) and the newer surface ionization techniques, including *desorption electrospray ionization* (DESI) and its derivatives, and *direct analysis in real time* (DART).

The ionization processes can create both positively and negatively charged ions (although not at the same time). Historically, positively charged ions (cations) were the first type studied, and they are still the type most commonly analyzed. Negatively charged ion (anion) formation is restricted to specific classes of compounds, including organic acids, nucleic acids, and halogenated environmental pollutants.

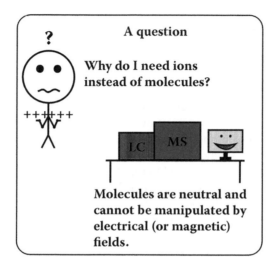

The amount of energy added during ionization and the structure of the analyte molecule determine whether the ions generated remain intact or fragment into smaller pieces. The various ionization methods supply vastly differing amounts of energy to the analyte, with EI being the most energetic, and therefore the most likely to cause fragmentation. Energetic ionization is termed hard ionization. Other approaches that preserve the molecular species intact are called soft ionization methods. The order of the extent of energy transfer from most to least is

$$EI > CI \approx APCI \approx APPI > MALDI \approx ESI$$

The same order applies to the breadth of compound classes to which an ionization method can be applied, particularly for small molecules. Thus, EI is an aggressive and widely applicable technique, whereas ESI ionizes efficiently only polar compounds and imparts little energy to the ions formed. ESI and MALDI are the favored ionization methods for large molecules. Note that MALDI is rarely used for molecules of <500 Da because of interferences resulting from the excess matrix with which a sample must be mixed prior to analysis (Section 2.2.4). The relationships among ionization methods, polarity, and molecular mass are illustrated in Figures 2.6 and 2.7.

As stated already, the amount of energy imparted to the neutral analyte molecules varies with the type of ionization and, combined with the structure of the

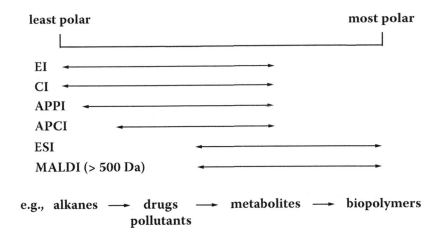

FIGURE 2.6 The choice of ionization method is often determined by the polarity of the analyte.

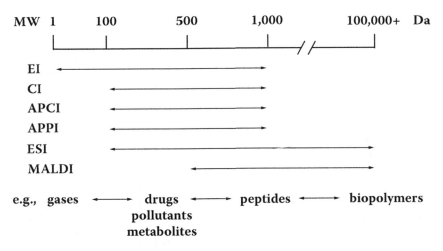

Several ionization methods are applicable for compounds
in the 100 to 1,000 Da range. The method chosen is often
determined by the nature of the objective, e.g., EI for
structural information and CI for quantification.
Above 1,000 Da, ESI or MALDI is usually selected.

FIGURE 2.7 Mass ranges for the different ionization methods.

analyte, determines whether fragmentation occurs. The different types of spectra
generated by EI, CI, APCI, and ESI are illustrated for di(ethylhexyl) phthalate
(a ubiquitous contaminant usually referred to as dioctylphthalate) in Figure 2.8.
In EI the molecular ion at *m/z* 390 is almost absent (~0.1% of the base peak) and
extensive fragmentation occurs, yielding major ions at *m/z* 167 and the base peak

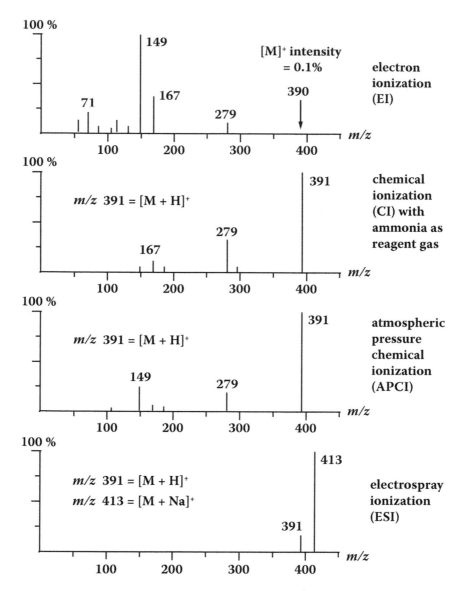

Spectra (simplified by removing the isotope peaks) illustrating how data varies, depending on the ionization method.

FIGURE 2.8   Mass spectra of dioctylphthalate under different ionization conditions.

m/z 149. The ion at *m/z* 149 is common to the phthalate family (it is the base peak in EI spectra) and is a useful indicator of whether an unknown contaminant is a phthalate. In CI, when ammonia is the reagent gas, the [M + H]⁺ ion forms the base peak of the spectrum along with some of the same fragments that are seen in the EI spectrum, although at much lower intensities. The APCI spectrum is similar to the CI spectrum. The $[M + H]^+$ ion is the most intense, and there are the familiar fragments at *m/z* 279, 167, and 149. The ESI spectrum illustrates the formation of alkali metal adduct ions, with $[M + Na]^+$ being the most intense species; $[M + H]^+$ also occurs. As expected, ESI is the gentlest ionization method, with no fragment ions observed.

2.2.1   ELECTRON IONIZATION (EI)

Electron ionization (previously called *electron impact* or *electron bombardment*) has a long history in MS, as it was the first widely used ionization method. EI sources, located inside the instrument's vacuum chamber, consist of a box (stainless steel, ~1 ml, also called the *ion volume*), with a series of openings that allow the introduction of both the sample and the ionizing electrons and the ejection of the resulting ions into the analyzer (Figure 2.9). EI sources are held at 200–250 °C to maintain the analyte(s) in the vapor phase and to prevent their deposition on the walls. The vapor pressure of the samples in the source must be in the 10^{-7}–10^{-2} Torr range. A heated tungsten or rhenium filament is used to produce an electron beam (*thermionic*

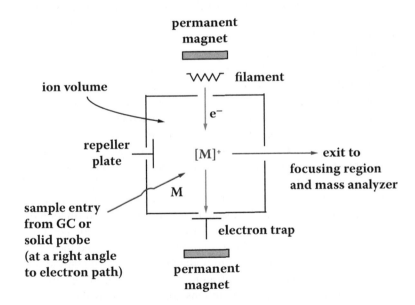

Electrons from the filament are collimated by a
magnetic field and traverse the source in a helical
path (to increase efficiency) to the electron trap.
The electrons react with the analyte molecules to
produce radical cations.

Reaction: $M + e^- = M^{+\cdot} + 2e^-$

FIGURE 2.9   Electron ionization (EI) source.

electrons) that traverses the source and bombards the analyte(s). A potential difference between the filament and a *trap*, on the opposite side of the source, directs the electrons across the ion volume. Small permanent magnets (located outside but adjacent to the source) collimate the electrons into helical paths as they traverse the source, thereby increasing the distance that the electrons travel and the likelihood of electrons ionizing the analyte molecules. Despite the increase in the path length of the electrons EI sources are inefficient, as only ~0.1% of the analytes are ionized. The current detected at the trap is used, in a feedback circuit, to control the electron current to about 200–300 μA. The (variable) potential drop between the filament and the trap, usually set to 70 eV (for both efficiency and reproducibility), provides energy to the electrons as they cross the ion source. A voltage on a *repeller* plate, located opposite the exit leading to the mass analyzer, acts to propel the ions out of the ion volume, perpendicularly to the electron stream, into a region where they are focused (with appropriate voltages on a set of ion lenses) and finally accelerated into the analyzer. The ion beam thus formed constitutes an electric current in the 10^{-7}–10^{-14} A range.

Ionization takes place when the energy acquired during the collision of a neutral molecule with an electron exceeds the ~10 eV, that is the typical ionization potential of small organic molecules. The removal of an electron from the highest occupied molecular orbital of the neutral molecule results in the formation of a positively charged ion. Because all electrons are paired in neutral molecules, the loss of an electron (which occurs preferably from a lone pair followed by a pi bonded pair and then a sigma bonded pair) results in an ion with an unpaired electron, a *radical cation*. The energy acquired is dissipated through vibration and electronic excitation, and when this can be accomplished without breaking a bond, then only the *molecular ion* is formed. For example, in aromatic compounds there is resonance among the bonds within the unsaturated rings that allows for the energy distribution that is necessary to preserve molecular ions. The EI spectrum of naphthalene, the simplest of the polynuclear aromatic hydrocarbons (PAHs), exhibits only the molecular ion, as illustrated in Figure 2.10a. However, there is usually a considerable excess of absorbed energy that leads to the breaking of bonds and to the formation of a variety of fragment and rearrangement ions. For example, in aliphatic compounds the bond structure reflects the saturated nature of the compounds, and the extra energy added during ionization cannot be accommodated when there are only single bonds between carbon atoms. The resulting fragmentation may be so extensive that no molecular ion remains and only fragment ions are observed, as shown for nonanol, where the mass of the highest *m/z* observed corresponds to the $[M - H_2O]^+$ fragment (Figure 2.10b).

An important advantage of EI spectra, obtained using 70 eV electrons (and keeping other experimental conditions constant), is that the fragmentation patterns, including relative intensities, are remarkably consistent and independent of the type of instrument used. The reproducibility of EI spectra arises from the fact that the energy of the ionizing electrons is so far in excess of that required to ionize the analytes that it overrides all effects attributable to other operational parameters and differences in ion source design. The rules governing the fragmentation of different organic compound classes are well established and documented. The

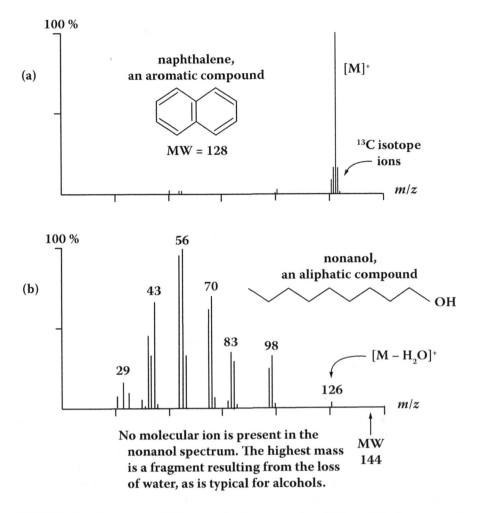

FIGURE 2.10   Comparison of EI spectra for (a) an aromatic and (b) an aliphatic compound.

reproducibility of EI spectra, both in terms of fragmentation and between instruments, has led to the creation of large libraries of spectra. The current NIST/EPA/ NIH library contains close to 250,000 EI spectra. Computers can be used to compare unidentified spectra to those in this library. The identity of the analyte is suggested where a match is found. However, care should be exercised in accepting computer-generated results by also considering whether the identity is commensurate with other available information. Examples of misassignments include highly unstable compounds found in environmental extracts or the presence of an unusual element in a biological sample.

Making life more complicated

How about a few more possible processes in EI:

Positive ions: dissociative ionization and dissociative rearrangements
Negative ions: electron capture and dissociative electron capture
Both polarities: ion-pair formation.

What are those 'lumps'
in the spectrum?

They are metastable ions that occur when 'excited species' leaving the
   source fragment in the flight tube. Indeed, they do look like lumps
   that are a few mass units wide.

The advantages of EI are that it is a universal, robust, energetic, and reproducible ionization method that provides structural information in the analysis of small molecules. The energy associated with EI is also a disadvantage because ionization often results in extensive fragmentation of the analyte frequently leading to the complete absence of the molecular ion, $[M]^+$. Despite not necessarily providing molecular mass information, EI may still be the method of choice to determine/confirm the identities of known compounds through library searches, or when aiming to obtain structural characteristics for unknown analytes. EI is most commonly used in GC-MS.

2.2.2 Soft Ionization

The ionization processes described in the following sections are classified as soft ionization methods. The intention is to reduce the amount of energy added to the analyte to the point where the only ions formed are directly related to the molecular mass and the energy is insufficient to cause the formation of fragment ions. To accomplish this aim, the amount of energy transferred to the analyte is reduced to the minimum required to ionize organic compounds, ~10 eV, much less than the 70 eV used in EI. CI is less efficient but more structurally selective than EI.

The predominant process of ion formation in the soft ionization methods is through the association of the analyte with an ionizing agent. Typically a proton (H^+) is attached to the analyte, yielding $[M + H]^+$ ions, but there are other species that can be adducted onto the analyte to form alternative ions in the molecular ion region.

Confusion in the molecular ion region

Is it a 'protonated molecule' or a 'protonated molecular ion'?

How many times do I have to say:
   It is a protonated molecule!
A protonated molecular ion would have two
   charges: the proton and the molecular ion.
By the way, avoid using 'quasi-molecular ion'
   and 'pseudomolecular ion'.

2.2.2.1 Chemical Ionization (CI)

In the 1960s, experiments on the ion-molecule chemistry of gases led to the development of CI, the first soft ionization method. In CI, the reduction of the amount of energy added to the analyte is accomplished by introducing into the ion source, along with the analyte, a vast excess of a *reagent gas*. The most common reagent gases are methane (CH_4), isobutane (C_4H_{10}), and ammonia (NH_3). An electron beam, similar to that used in EI, provides the initial ionization of the reagent molecules by a process similar to that in EI. The resulting ions from the reagent gas then react with other neutral reagent gas molecules to form proton-donating species, such as CH_5^+, $C_4H_9^+$, and NH_4^+, which are characteristic of the CI process. In turn, these charged species (ions) collide with the analyte molecules and transfer a proton, in ion-molecule reactions, to form protonated species, i.e., $[M + H]^+$.

The design of CI sources is based on the need to achieve a reagent gas pressure of about 1 Torr (1.3 mbar). Accordingly, CI sources must be almost gas tight, much more enclosed than EI sources (Figure 2.11). This is accomplished by making the exit slit to the analyzer smaller than in EI sources and by removing the electron trap. The latter is not needed because the reagent gas not only reacts with the electrons but also acts as a barrier preventing the electrons from traversing the source to

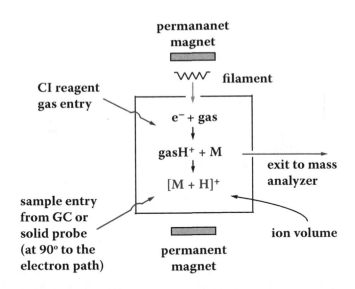

Electrons from the filament react with a reagent gas (methane, isobutane, or ammonia) generating protonated reagent species that, in turn, transfer a proton onto, or form an adduct with, the analyte.

$$e.g., M + [NH_4]^+ \rightarrow [M + H]^+ \text{ and/or } [M + NH_4]^+$$

FIGURE 2.11 Chemical ionization (CI) source.

the trap. Similarly, there may be no need for a repeller because the flow of gas out the source exit slit can be used to transfer the ions out of the ionization volume and into the focusing and acceleration regions located prior to the mass analyzer.

The efficiency of CI is enhanced by raising the energy of the electron beam to 200 eV thereby increasing the formation of the reagent species. Also, the vacuum system must include differential pumping (Section 2.5) to remove the excess reagent gas so that it does not affect the pressure in the mass analyzer.

Another way to reduce the amount of energy added to the analyte ions is to operate CI sources at a temperature of ~150 °C, rather than the 250 °C used with EI sources. The lower source temperature can significantly reduce the extent of fragmentation of some compounds, while the presence of the reagent gas acts as a buffer, minimizing the likelihood that analyte ions will be lost by condensation onto the cooler source walls.

Given that the primary objective of CI is to transfer a proton from the reagent ions onto the analyte, the reactivity of the donating ions is of interest because it affects both the amount of energy transferred (therefore the extent of fragmentation) and the efficiency of the ionization process. The reagent ions formed from methane (e.g., CH_5^+ and $C_2H_5^+$) have the highest energies among the common reagent gases, and thus transfer the most energy to the analyte. The amount of energy imparted is reduced from methane to isobutane to ammonia.

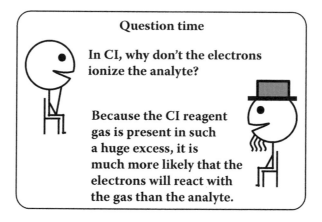

Question time

In CI, why don't the electrons ionize the analyte?

Because the CI reagent gas is present in such a huge excess, it is much more likely that the electrons will react with the gas than the analyte.

An alternative way to consider the reactivity of the reagent ions involves proton affinity, a measure of the strength of the bond between the reagent gas and the ionizing protons. Values of proton affinities are in the reverse order to the amount of energy transferred to the analyte. For example, CH_4 has a lower proton affinity (5.7 eV) than that of the other reagent gases and, consequently, CH_5^+ will readily give up the excess proton to an analyte. On the other hand, NH_4^+ is stable when compared to CH_5^+ because of the high proton affinity of ammonia (8.9 eV), making NH_4^+ a poor reagent ion for compounds with low proton affinities, e.g., saturated hydrocarbons. A practical application of the fact that NH_4^+ does not ionize hydrocarbons is the use of ammonia as the reagent gas to minimize the observation of alkane contaminants.

Another ionization process that competes with proton transfer during CI is the reaction of reagent gas ions with the analyte to form adducted ions that occur at m/z values higher than that of the protonated molecule. For example, methane may yield $[M + C_2H_5]^+$ and $[M + C_3H_7]^+$ ions, isobutane $[M + C_4H_9]^+$, and ammonia $[M + NH_4]^+$ and $[M + N_2H_7]^+$. With ammonia as the reagent gas, the $[M + NH_4]^+$ adduct ions are often the dominant species. Adduct ions may be utilized to differentiate ions that are derived from intact molecules from those that are fragments because the latter ions rarely have adducted species; e.g., with methane, both $[M + H]^+$ and $[M + C_2H_7]^+$ (M + 1 and M + 29, respectively) ions can occur, while for ammonia the $[M + H]^+$ and $[M + NH_4]^+$ (M + 1 and M + 18, respectively) ions are often seen.

The choice of reagent gas is very important as each has advantages and drawbacks. Methane is universally applicable and is the most commonly used gas, but it imparts the most energy to the analytes that may result in extensive fragmentation. Isobutane can ionize even the most nonpolar compounds, e.g., the alkanes, while imparting little energy, and therefore maintaining strong signals for the molecular species. A drawback of isobutane is that it polymerizes during the ionization process, resulting in deposits that cause short circuits necessitating frequent cleaning of the source. Ammonia is arguably the most useful and convenient reagent gas as long as molecular mass information/quantification is the primary objective and alkanes are not the subject of the analysis. Also, ion sources require less frequent cleaning when ammonia is used.

Making life complicated

There is a confusing process that can occur with ammonia CI.

Some spectra may appear to show $[M]^+$ and $[M + NH_4]^+$, i.e., (M and M + 18).

In fact, the apparent $[M]^+$ ion is actually $[M + NH^4 - H_2O]^+$, i.e., (M + 18 − 18).

The addition of ammonium and loss of water can be shown readily by accurate mass measurement.

A particular advantage of CI for quantitative analyses is that the total ion current (TIC) is not a composite of the molecular species and fragments but is composed only of ions from the molecular ion region; this is ideal for selected ion monitoring and selected reaction monitoring (Section 3.3.3.1). Furthermore, because molecular species are produced predominantly in CI, the likelihood of cross-contamination (unwanted contributions to the m/z of the analyte) is reduced by the lessening of the number of fragments from other components of the sample.

Negative chemical ionization (NCI) is a misnomer because only rarely is there an actual chemical reaction involved. Instead, the process requires *thermal electrons* that are created when a gas in the ion source is used as a buffer to decelerate and thereby reduce the energy of electrons arriving from the filament. Methane is often used as the buffer gas. The structure of methane makes it a more effective agent for absorbing energy from the electrons than are inorganic gases, such as nitrogen, although the latter can also be used. The energy of the electrons is reduced from 70 eV to <1 eV, at which point they can be captured by analytes with high electron affinities. Typical analytes include halogenated compounds, such as polychlorinated biphenyls (PCBs). When applicable, the sensitivities obtained with NCI may be orders of magnitude higher than those attainable in positive ion mode using either EI or CI. Such sensitivities, however, depend on the structure of the analyte, including the number and positions of chlorine and fluorine atoms present. Responses for brominated compounds are poor in the molecular ion region, as the acquisition of an electron results in fragmentation and the formation of a $[Br]^-$ ion.

Names that attempt to better describe the ionization process include *electron capture negative ionization* (ECNI) and *electron resonance capture negative ionization*.

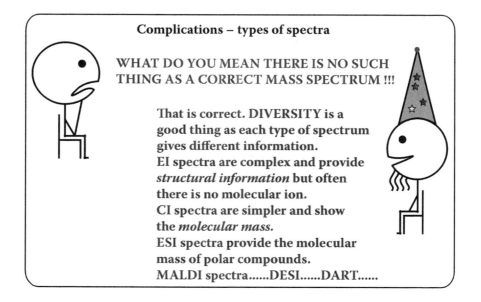

Complications – types of spectra

WHAT DO YOU MEAN THERE IS NO SUCH THING AS A CORRECT MASS SPECTRUM !!!

That is correct. DIVERSITY is a good thing as each type of spectrum gives different information.
EI spectra are complex and provide *structural information* but often there is no molecular ion.
CI spectra are simpler and show the *molecular mass*.
ESI spectra provide the molecular mass of polar compounds.
MALDI spectra......DESI......DART......

2.2.2.2 Electrospray Ionization (ESI)

ESI (and MALDI; Section 2.2.4) has revolutionized mass spectrometry since the 1990s, and it is now the dominant ionization method for polar compounds, from small molecules to biopolymers, opening up vast new areas of research that were hitherto inaccessible to mass spectrometry. ESI is the softest ionization technique; the energies involved are barely above those necessary to generate ions.

ESI has overcome a major problem that frustrated mass spectrometrists for many years: how to volatilize polar compounds without also causing decomposition. Samples used to be introduced into the ion source by evaporation from a solid probe. Heating the probe often failed to produce useful spectra for polar compounds, resulting instead in brown residues of decomposed material on the probe tip. The ESI process does not vaporize the analytes from a surface; instead, ions are formed in a charged, nebulized solution of the analyte. As the droplets are dried (desolvated), they disintegrate, releasing ionized analytes into the gas phase. Furthermore, the evaporation takes place at atmospheric pressure, outside the vacuum chamber, and almost all gases derived from the liquid flow are removed outside the instrument, thereby maintaining the vacuum inside the mass analyzer.

In ESI, analytes, dissolved in an electrically conductive liquid medium, flow through a narrow steel tube where they are subjected to a high voltage, 2–5 kV (positive or negative), with respect to the vacuum chamber. The voltage removes ions of polarity opposite to that of the applied voltage, yielding a solution where there is only one charge type; this is essential to the ionization process. There is an outer tube, coaxial with the tube carrying the liquid, through which there is a nitrogen flow (5×10^3 Torr, ~7 bar, 100 psi) acting as a sheath gas that nebulizes the eluate into a spray of charged droplets that are then dried (desolvated), e.g., with a secondary flow of heated nitrogen (Figure 2.12).

The evaporation progressively forms smaller and smaller droplets with the eventual release of ions into the vapor phase. The desolvation causes the density of the electric field of the droplets to increase to the point where disintegration occurs. There are two models for the disintegration, *Coulombic* and *jet* fissions. When the repulsive Coulombic forces between the like charges on the surface of the droplets exceed the forces attributable to surface tension (the Rayleigh instability limit), the

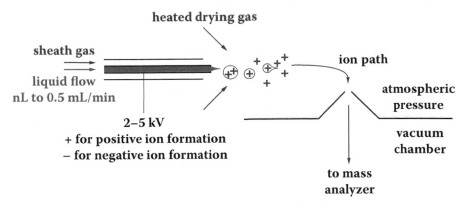

Two concentric steel tubes carry the liquid and nebulizing gas.
After nebulization, additional gas is used to dry the droplets.
As the droplets shrink the charge density increases to the point at
   which the droplets are no longer stable and ions are released
   from the liquid matrix into the vapor phase.
Ions enter the mass analyzer at an angle to reduce the entrainment
   of neutral molecules.
At low flow rates, such as in nanospray ESI, spontaneous
   evaporation is sufficient to eliminate the need for the drying gas.

FIGURE 2.12 Electrospray ionization (ESI) source.

droplets disintegrate into smaller units which still remain charged. When the droplets are reduced to a radius of ~10 nm, the electric field begins to support direct ion evaporation. During this type of evaporation the droplets distort and develop an extension with concave surfaces, called a Taylor cone, from which the ions are released (Figure 2.13). The figure also illustrates the development of another Taylor cone, this time at the tip of the ESI probe where the droplets are formed as the solvent flow is nebulized. Taylor cones form as a consequence of the interaction of surface tension and charge repulsion. Formation of the Taylor cone at the tip of the ESI probe is important because it results in a stable stream of droplets from which the ions are eventually released.

The direction of the flow of the ions produced in ESI (as well as other API methods) is usually at an angle to the entrance into the mass analyzer. Under the combined influence of a voltage gradient, and the vacuum behind the entrance, the ions deviate from the angle at which they originated (from the ESI probe) and enter the

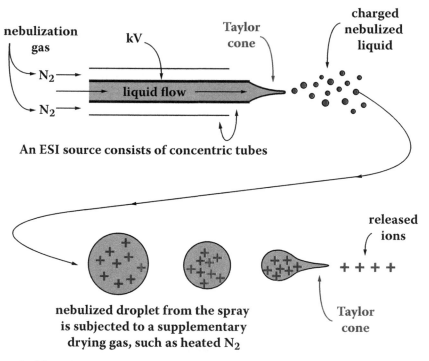

Stable nebulization of the liquid flow occurs when a Taylor cone is established at the tip of the ESI source.
As the droplets are dried, the charge density increases, resulting in droplet disintegration either because of Coulombic forces or by distortion to form a Taylor cone from which ions are released.

FIGURE 2.13 Taylor cone formation occurs both at the end of the ESI tube and as the droplets disintegrate to release ions.

analyzer. An angle of 90° is shown in Figure 2.12, but instrument manufacturers employ various angles (which may be adjustable). The movement of the ions away from the stream of neutral molecules reduces the amount of nebulizing gas and solvent vapor introduced into the vacuum chamber, lessening the background noise. Despite the removal of most neutral molecules in the exhaust from the ion source, a mixture of disintegrating droplets, ions, and nebulizing/drying gas still enters the evacuated section of the instrument through a series of chambers that are maintained at progressively higher vacuum. Differential pumping removes the excess gas and solvent molecules before the ions enter the mass analyzer.

As noted above, the solutions must be electrically conductive, thus the acetonitrile:water and methanol:water mixtures used for reversed-phase LC are appropriate. The addition of a little acid (0.1% acetic or formic acid) to the solvent enhances chromatographic separation and also supplies protons for the ionization process. Trifluoroacetic acid (TFA), the acid most commonly used in reversed-phase LC (when detection is by UV absorption), is not used in LC-ESI-MS, as this acid compromises ion formation because of the strong ion pair formation between the analytes and TFA. Other buffering solvent systems, aiming to improve chromatographic performance, must include volatile reagents, such as ammonium acetate, instead of involatile species such as sodium and potassium phosphates. When sample is flowing directly into an ESI source without chromatographic separation (flow injection analysis), other solvents e.g., 100% methanol may be used as long as they are conductive.

The combination of nebulizing and desolvation gases (probe parameters) described above is suited to liquid flows of ~0.2 ml/min. Because mass spectrometry is a *concentration-dependent* technique, reducing solvent flow will increase the residence time of the analyte in the source. The consequent enhancement of absolute sensitivity can reach 1,000-fold. The potential for such increases in sensitivity led to the development of instrumentation and techniques for LC where solvent flows are 0.3–1.0 µl/min, nano-LC. At such flow rates fused silica and stainless steel lines (with tips tapered to ~10–20 µ) can be used as the ESI source. The flow rates in nano-LC also permit the reduction of the nebulization gas pressure to ~250 Torr (<1 bar), sometimes even to zero. Furthermore, a drying gas is no longer required. When there is no need to separate the components of the sample, small volumes of material (~1 µl) can be loaded into a capillary and placed in the ESI housing. In this case no gases are needed and the potential difference between the liquid in the capillary and that in the instrument chamber is sufficient to create an *electro-osmotic* flow of nebulized droplets from the tip of the capillary.

In positive ESI, ions are formed from neutral analytes by protonation or other cation attachment. The charge is usually provided by addition of a proton to yield $[M + H]^+$ ions. However, alkali metal cations (particularly sodium and potassium), and also the ammonium cation, can be added to analytes to form $[M + Na]^+$, $[M + K]^+$, and $[M + NH_4]^+$ ions, respectively. Other alkali metals ions, such as lithium and even cesium, may also serve as the ionizing species. In all these instances a charge is being adducted onto the molecule; but the term *adduct ion* is used for species other than the protonated form.

In the negative ion mode, deprotonation or anion attachment occurs. Carboxylic acids are often the target molecules, with ions forming through the loss of a proton, $[M - H]^-$. The addition of a negatively charged species, such as from a halogen or an acetate anion, can also occur, yielding, for instance, $[M + Cl]^-$ and $[M + CH_3COO]^-$ ions. Negative ESI is also applicable for other species that can give up a proton readily, such as (covalent) sulfates and phosphates, with the latter making this ionization mode suited to nucleic acids.

Because of the limited amount of energy acquired by the analytes, the mass spectra produced by ESI are dominated by molecular species. The interpretation of the spectra is straightforward because there is usually no fragmentation. The apparently simple process of forming ions (though the true mechanism of ESI remains a matter of debate) is both a strength and a weakness. A strength is that there is no active chemical process involved, such as bombardment with electrons or reaction with a proton donating species, as in EI and CI, respectively. Indeed, ESI is based on just the propensity of the analyte to form ions in a solution to which a voltage has been applied. The result is a very gentle ionization method that almost always provides information on the molecular mass of the analyte. This strength is also a weakness because it makes ESI highly dependent on the polarity of the analyte, i.e., its ability to form ions while in solution by acquiring a proton or alkali metal ion. Take, for instance, a mixture consisting of three compounds: a steroid with a single hydroxyl group, a natural product containing multiple amines, and the tetrabutylammonium cation. The nonpolar/lipophilic steroid will give no or very poor ESI response, while the natural product will ionize readily. However, the most intense response will occur for tetrabutylammonium, which is ideally suited to ESI because it is already a cation. The more polar the analyte, the lower will be its limit of detection; thus, a relatively small amount of a polar species in a mixture of compounds can dominate the spectrum, giving a misleading indication of the true composition of the sample.

The range of concentrations (dynamic range) over which ESI can be used is only three to four orders of magnitude, after which the ionization process saturates. Therefore, ESI can be thought of as a method in which the amount of ion current that can be produced is limited irrespective of how much material is introduced. The lack of any separation in the flow injection mode makes the problem of selective ionization particularly problematic, as the most polar species will sequester the ionization capacity when mixtures are analyzed. The limited capacity for ionization is less of a problem when LC is used to separate compounds with different polarities, as each compound is ionized individually upon elution from the column.

A unique characteristic of ESI is the propensity to form multiply charged ions, i.e., $[M + nH]^{n+}$, from large molecules. Numerous multiple charge states may occur, generating an *envelope* of ions. Because mass analyzers determine *m/z* ratios rather than mass, multiple charging leads to ions with *m/z* values that make it possible to analyze molecules of high molecular mass routinely (up to 150 kDa) with simple and inexpensive mass analyzers, e.g., quadrupoles, that have a mass range of only 4 kDa. For example, a peptide with a molecular mass of 5 kDa that forms ions with two, three, and four charges will yield *m/z* 2,501, 1,667.7, and 1,251, respectively, i.e., $(5{,}000 + 2)/2$, $(5{,}000 + 3)/3$, and $(5{,}000 + 4)/4$. These *m/z* values can be readily observed using an instrument with a mass range of only 3 kDa.

When sufficient resolution is available, i.e., with TOF and FT systems, the spacing of the ions in the isotope pattern will show the charge state (i.e., the number of charges on the ions). In the example above, the spacing on the m/z x-axis would be 0.5, 0.33, and 0.25 Da for the two, three, and four charge states, respectively. The charge states are usually too high for the direct determination of the isotope spacings of proteins, except when FT-ICRMS is used. Commercial software packages are available for the deconvolution of the measured raw (m/z) data of multiply charged ion envelopes into the average (zero charge) mass of the analyte by solving a set of simultaneous equations.

Capillary electrophoresis and mass spectrometry can be coupled using a special ESI interface that provides an additional makeup flow, comprised of water, organic solvent, and acid, to create suitable conditions for ESI (Figure 2.14). Flow rates and sample loading are very low in CE. The supplemental sheath makeup liquid, while useful in the ESI process, has the effect of diluting the analyte, thereby compounding the sensitivity problem caused by the small amounts of analyte used in CE. The importance of CE is that the very large number of theoretical plates available enable complex separations that can reveal subtle differences in analytes.

2.2.2.3 Atmospheric Pressure Chemical Ionization (APCI)

APCI is the equivalent of CI conducted at atmospheric pressure, which is $\sim 10^3$ higher than that in a conventional CI source. The filament used in CI to produce the

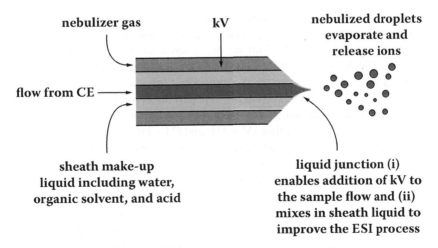

A modified electrospray interface is used to couple
   CE to MS.
A secondary sheath liquid, to which a kV potential is
   applied, includes organic solvent, water, and acid to
   facilitate the ESI process.
Although necessary for the ESI, the sheath liquid
   dilutes the CE flow and reduces sensitivity.

FIGURE 2.14 An ESI interface for capillary electrophoresis.

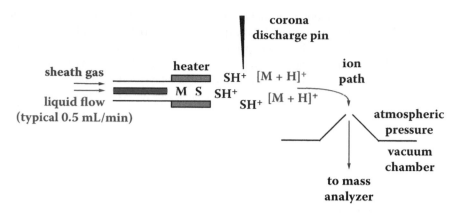

Evaporated solvent (S) is ionized and becomes a proton donating
species, SH$^+$, in the plasma derived from the corona discharge.
The protonated solvent acts the same way as the reagent gas in CI,
transfering protons to the analyte, M, to yield [M + H]$^+$ ions.
This method is used for compounds with masses <1 kDa.

FIGURE 2.15   Atmospheric pressure chemical ionization (APCI) source.

electrons that ionize the reagent gas, would burn out immediately at such pressures.
Instead, a high voltage (~3 kV) is placed on a corona discharge pin (needle) situated
in the ion source (Figure 2.15). Samples are sprayed into the source as LC (or GC)
effluents (or injected directly as solutions) and are evaporated by heating and adding
a gas, usually nitrogen, to sweep the resulting vapor into the ionization region. A discharge occurs as the *plume* of solvent is swept past the corona pin, creating a plasma
in which both positive (protonated) solvent ions and electrons are generated that, in
turn, ionize the analytes, forming [M + H]$^+$ and [M – H]$^-$ ions. Other ions, such as
[M]$^+$ and [M]$^-$, can also form because the ionization process consists of a highly
complex set of ion-molecule reactions involving proton transfer, electron capture,
charge transfer and exchange, clusterings, and substitutions. As with ESI, the stream
of ions formed is set at an angle to the entrance of the mass analyzer to lessen the
entrainment of neutral molecules and reduce the background signal.

APCI is a more energetic process than ESI because of both the chemical reactions
involved and the requirement for desolvation temperatures to be in the 200–500°C
range. These conditions limit APCI to small molecules (<1 kDa) with intermediate polarity, i.e., to compounds that can be protonated readily and do not disintegrate when exposed to the heat necessary to remove the solvent. The technique falls
between EI and ESI with respect to the classes of compounds to which it can be
applied; e.g., steroids are readily accessible to APCI but are often transparent in ESI
because of their low polarity. Nonpolar compounds, e.g., alkanes and polynuclear
aromatic hydrocarbons, can be analyzed by EI (and CI) but not APCI.

The advantages of APCI are that its ionization efficiency is more uniform than
that of ESI as it is not as affected by variations in the polarity of the analytes. The

reduced structure dependence means that APCI is useful when searching for and quantifying members of entire families of compounds, e.g., a drug and its metabolites. Accordingly, APCI is often used in the pharmaceutical industry to investigate the fate, distribution, and kinetics of drugs in biological systems.

I decided to lose an electron and fly solo today. The question is whether I will be stable with all that extra energy or, in the end, will I just be a bunch of fragments. It sure will be a thrill.

Not me, I am linking up with my low energy friend H⁺ who has promised to make me very stable. I really want to know who I am at the end of the ride.

2.2.2.4 Atmospheric Pressure Photo-Ionization (APPI)

APPI is carried out in ion sources essentially the same as those used in APCI, except for having an ultraviolet (UV) light in place of the corona discharge needle (Figure 2.16). An important aspect of APPI is that the energy provided by the UV light is ~10 eV, just above that required to ionize most organic compounds but below that of the water, methanol, and acetonitrile used in reversed-phase LC, as well as below the ionization energies of atmospheric gases such as nitrogen. Consequently, these solvents and gases do not contribute to the background signal.

There are two competing ionization processes in APPI. The first is brought about when the energy of the UV radiation is sufficient to eject an electron from the analyte to form $[M]^+$ ions, e.g., the polynuclear aromatic hydrocarbons (PAHs), where the aromaticity is ideal for absorbing the UV light, and the acquired energy is sufficient to expel an electron. The second mechanism is utilized for analytes that cannot be ionized directly by UV light. In this case a dopant, e.g., toluene or acetone, is added to the vapor stream passing the UV source. As with the reagent gas in CI, a series of reactions is engendered by the UV light resulting in the formation of $[dopant + H]^+$ and $[solvent + H]^+$ species that act as reactive proton donors transferring the proton to form $[M + H]^+$ ions of the analytes.

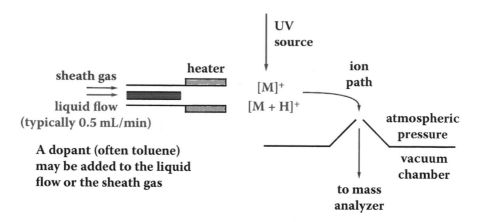

Ionization is mediated by UV light directly or via a dopant.
Direct ionization by UV light leads to [M]$^+$ ions.
With an added dopant a CI type process gives [M + H]$^+$ ions.
Both positive and negative ions can be formed.
APPI expands the API methods to less polar compounds,
  e.g., aromatic hydrocarbons.
This method is used for masses <1 kDa.

FIGURE 2.16   Atmospheric pressure photo-ionization (APPI) source.

APPI is useful because it overlaps with APCI for some compounds while extending the range of potential analytes to less polar species, e.g., PAH. APPI also reduces the background signal from LC solvents.

2.2.3 SURFACE IONIZATION METHODS

A series of technologies has been developed during the last 5 years (with an alphabet soup of acronyms) for the identification of compounds adsorbed on surfaces. The most common methods are *desorption electrospray ionization* (DESI) (and related techniques) and *direct analysis in real time* (DART). Applications range from the identification of explosives and drugs in forensics to biological arenas, such as the characterization of tissue sections.

2.2.3.1 Desorption Electrospray Ionization (DESI)

A DESI source is similar to an ESI source except that there is no secondary flow of heated nitrogen to dry the nebulized droplets (Figure 2.17), so a spray is produced. A mixture of methanol and water is subjected to a potential of 3–5 kV and then nebulized with nitrogen supplied at ~5 × 10^3 Torr (~7 bar). The spray mixture may be modified with acid (acetic or formic) or ammonia to enhance ion formation. The stream of charged droplets is then directed at the surface of solid samples, a few millimeters away, from which the analyte(s) is ejected. The desorbed compounds leave

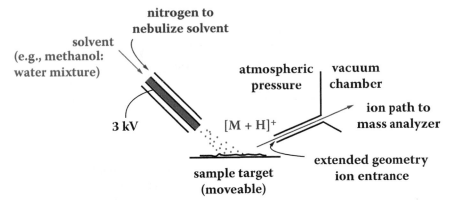

Charged solvent droplets are directed at a surface, and protonated analyte, [M + H]⁺, ions are desorbed.
[M − H]⁻ ions form when ammonia, a basic modifier, is used.

FIGURE 2.17   Desorption electrospray ionization (DESI) source.

the surface as ions or are ionized in the resulting plume. The species formed are [M + H]⁺ ions or [M − H]⁻, when ammonia is used as a modifier.

The selection of the angle of incidence of the droplets, i.e., the angle at which the droplets arrive at the sample surface, is important and depends on the molecular mass of the analyte, e.g., ~40° for small molecules and ~75° for peptides and proteins. Recovery of the ions formed occurs at an angle of ~10°, a little more for small molecules and a little less for biopolymers. Ions enter the mass analyzer through a collection tube that is extended close to the surface analyzed (Figure 2.17).

The DESI mass spectra of biopolymers include multiply charged ions and resemble those obtained in ESI. Ionization efficiency depends on several experimental parameters related to surface effects, the chemical composition and flow rate of the spraying solvent, the applied voltage, and importantly, the angle of the spray.

Several other ionization methods have been developed based on DESI, including *desorption atmospheric pressure chemical ionization* (DAPCI), *desorption atmospheric pressure photo-ionization* (DAPPI), *laser ablation electrospray ionization* (LAESI), and *extractive electrospray ionization* (EESI). Each technique uses variations of the solvent, how the charged beam is formed, and how the beam is used to facilitate the production of analyte ions. Because these are surface methods (except EESI), they are incompatible with LC.

2.2.3.2   Direct Analysis in Real Time (DART)

DART is different from the other surface desorption ionization methods in that an excited gas stream, typically helium (sometimes N₂ or Ne), is used to ionize the analytes (Figure 2.18). The neutral gas atoms (or molecules), flowing at low velocity (1–5 L/min), are subjected to a corona discharge that creates a mixture of ions, electrons, and excited (metastable) gas atoms (or molecules). All charged species are

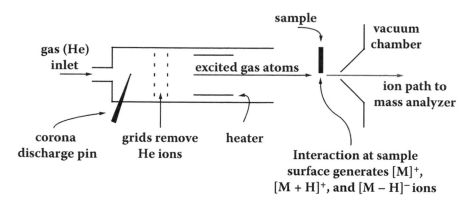

A gas (usually helium) flows through a corona discharge where it
  can be ionized or form excited (metastable) atoms.
The ions are removed from the gas stream and the remaining
  excited atoms, heated as necessary, are directed onto the sample.
Energetic atoms generate ions directly at the sample surface
  by causing ejection of an electron to form $[M]^+$ ions, or
  alternatively, ambient water is ionized and forms a cluster with the
  analyte where proton transfer yields $[M + H]^+$ ions.
A competing process that ionizes oxygen leads to negatively
  charged species, e.g., $[M - H]^-$ ions.
Heat from the helium flow, combined with like-charge repulsion of
  ionized sample, evaporate the analyte ions. The desorbed ions
  enter the mass analyzer via a skimmer cone.

FIGURE 2.18   Direct analysis in real time (DART) source.

removed by passage through a set of grids, with opposite polarities, and the resulting
stream of energetic gas atoms (molecules) is used to vaporize and ionize the analytes
on the target surface. The excited gas stream may be heated to 550°C to assist in the
vaporization of the samples. The method can be used to produce either positive or
negative ions. Ions enter the mass analyzer via skimmer cones similar to those used
in the other atmospheric ionization methods.

There are competing ionization processes in DART. If the energy transmitted
to the analyte upon collision with the excited gas atoms (molecules) exceeds its
ionization energy, an electron can be ejected forming $[M]^+$ ions. Alternatively, the
energized gas may interact with water in the ambient atmosphere, forming a proton
donating species that in turn protonates the analyte yielding $[M + H]^+$ ions. Negative
ions may be formed in multiple ways that include proton abstraction or the capture of
thermalized electrons emitted from surfaces with which the excited gas species have
interacted. The thermal electrons may also react with ambient oxygen, yielding a
negatively charged oxygen radical ($O_2^{-\bullet}$). Analytes that react with the oxygen radical
form various negative ions, including $[M - H]^-$, $[M]^-$, and $[M + O_2]^-$.

DART is a *line-of-sight* technique, in which the stream of metastable gas is in line with the entrance into the mass analyzer. As a result, the method is most effective when the analyte is located on the edge of an object that is then brought up to the gas stream so that the ionization and evaporation do not disrupt the flow toward the entrance of the vacuum chamber.

DART has been used to vaporize and ionize a large variety of analytes, e.g., drugs, explosive residues, plant metabolites, peptides, and oligosaccharides, from a variety of surfaces, including glass, paper, fruits, fabrics, medicinal tablets, and thin-layer chromatography plates. Limitations of DART include that it is useful only for small molecule analysis and cannot be combined with chromatography because it is a surface ionization method.

DART is easy to use and rapid. To release ions from the surface simply requires placing the samples in the flow of the excited gas atoms. As an example, there is an often heard rumor that U.S. dollar bills that have been in circulation will inevitably be contaminated with cocaine. One of the authors (J.G.) placed a $1 bill in a DART source linked to a time-of-flight analyzer. An ion was immediately observed that was within 5 ppm of *m/z* 304.1585 Da corresponding to the [M + H]+ of cocaine.

2.2.4 MATRIX-ASSISTED LASER DESORPTION/IONIZATION (MALDI)

Lasers have been used for many years, with varying degrees of success, to ionize molecules directly. The energy added was usually excessive and resulted in the destruction or extensive fragmentation of the analytes. The significant conceptual differences between direct laser ionization and MALDI are that in the latter there is a matrix present, usually an aromatic acid, in vast molar excess over the analyte and that the matrix absorbs most of the energy supplied by the laser.

The matrix and analyte solutions are co-deposited onto special plates (steel, sometimes gold coated) where hundreds of sample "spots" can be placed. Evaporation of

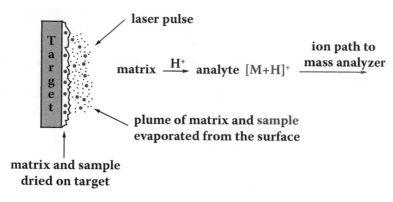

FIGURE 2.19   Matrix-assisted laser desorption/ionization (MALDI).

the solvent leads to the formation of crystals in which the analytes are embedded and co-crystallized. Upon placing the sample plate into the vacuum chamber, the components of these crystals are transferred into the vapor phase by the energy absorbed on exposure to a pulsed (UV) laser (Figure 2.19). The excess, as well as the aromaticity, of the matrix provides for the absorption of most of the energy from the laser. The absorbed energy excites the matrix molecules, and the process (not fully understood) results in the formation of protonated matrix ions that have proton affinities lower than those of the analytes. The laser adds excess energy very rapidly (nanoseconds) to the sample, raising the temperature to about 1,000 °C, when explosive evaporation occurs. The speed of the process reduces the probability of degrading the analyte. The evaporation also carries neutral analyte molecules into the vapor phase where charge transfer takes place from the protonated matrix to any compound with a higher proton affinity, yielding protonated molecules, $[M + H]^+$. Analyte ions may also be generated by adduct formation, e.g., $[M + Na]^+$, $[M + K]^+$, depending on the availability of alkali metals from the sample or from the original solution.

The selection of the matrix is of major importance in MALDI because the amount of energy transferred to the analyte is matrix-specific. Examples of matrices include dihydroxybenzoic acid, alpha-cyano-4-hydroxy-cinnamic acid, and sinapinic acid (also called sinapic acid). "Alpha-cyano" is the choice for peptides up to 5 kDa, while sinapinic acid is used for larger peptides and proteins. The reason for the changeover from one matrix to the other is that alpha-cyano transfers larger amounts of energy to analytes than does sinapinic acid, resulting in the destruction of larger peptides and proteins. Samples and matrices are preferably prepared in the same solvents, e.g., an acetonitrile:water (1:1 v/v) mixture that may contain 0.1–1% trifluoroacetic acid. The concentration of the matrix is ~10 mg/ml. The simplest approach with alpha-cyano and sinapinic acid, and an acetonitrile:water solvent mixture, is to prepare saturated solutions. Alpha-cyano should be prepared daily and protected from light. Other matrices are stable and can be stored for several days.

There are multiple ways to "spot" samples. For instance, the sample and matrix may be mixed before spotting, or either solution may be spotted first with or without drying in-between. Often it is worth trying multiple approaches to sample loading to see which provides the best signal-to-noise ratio. Robotic placement of samples is applicable for the eluates from nano-LC. The signals from automated, sequential analysis of hundreds of spots replicate LC chromatograms. In these applications the MALDI plate is often pretreated with the matrix solution to provide an appropriate surface for the analytes.

Assuming that the analyte is a single compound (or a mixture of no more than three or four compounds) with concentrations in the pmol/μl range, e.g., isolated by chromatography or gel electrophoresis, mixing the analytes with a saturated matrix solution will give a molar ratio of ~1:1000. A typical 0.5–2 μl spot on a MALDI plate will thus contain ~20 nanomoles of matrix and a few picomoles of analyte.

MALDI is similar to CI in that a medium, present in excess, is ionized (absorbing most of the imparted energy) and then transfers an ionizing species, typically a proton, to the compound of interest. The transfer is a low-energy process that results in minimal fragmentation, even for proteins (Figure 2.20). MALDI spectra are composed predominantly of singly charged ions, although double and triple charge states may be seen when analyzing species with larger molecular mass, as illustrated by the low intensity $[M + 2H]^{2+}$ ion of myoglobin in Figure 2.20.

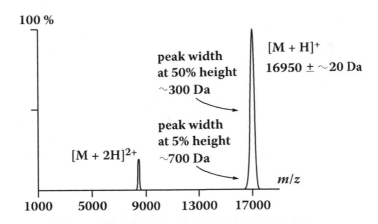

Holo form of myoglobin analyzed. Sample preparation
   denatures the protein so the noncovalent heme group
   (616 Da) is lost and the observed mass is for apo-myoglobin.
Spectra consist primarily of $[M + H]^+$ ions. Ions with higher
   charge states, usually $[M + 2H]^{2+}$, have low intensities.
Accuracy of mass determination for proteins is ±0.1%.
Resolution is limited; therefore, peaks are relatively broad.

FIGURE 2.20   MALDI-TOF spectrum of myoglobin, a protein.

Nd:YAG lasers (frequency multiplied to provide a 355 nm output) predominate in modern instruments because of the high repetition rates available and their long lifetimes. Nitrogen (337 nm) lasers are also used in older instruments. MALDI sources are usually coupled with TOF analyzers because of the availability of the mass range necessary to observe singly charged proteins. MALDI is an in vacuo technique, although atmospheric pressure MALDI does exist.

MALDI has been employed extensively in a wide variety of proteomic studies. The technique has also been applied to the analysis of other biopolymers, such as nucleic acids, although care must be taken to avoid sodium or other alkali metal ions in the sample or matrix because varying amounts of these species will be absorbed onto the nucleotide phosphate groups, causing peak broadening. Methods to remove alkali metal ions include adsorbing them onto ion exchange resins, displacement with ammonium, and observing negative ions where $[M - H]^-$ ions predominate. Specific matrices for nucleic acid analysis include trihydroxyacetophenone (>15 kDa) and (<15 kDa) 3-hydroxypicolinic acid.

Synthetic polymers are a set of individual compounds with chain lengths that differ from each other by the mass of the monomer, e.g., 44 Da ($-CH_2CH_2O-$) for polyethylene glycol. Such polymers can also be analyzed by MALDI, provided they exhibit sufficient polarity so that a charge can be attached. Dithranol is the matrix often used; including a silver salt, such as silver trifluoroacetate (0.1 mg/ml), in the matrix may facilitate cation attachment. Hydrocarbons, such as polyethylene, cannot be ionized effectively because they lack sufficient polarity for proton attachment. Because multiple compounds are being analyzed at the same time, the concentration of the analyte often has to be increased; this reduces the analyte:matrix ratio and compromises the MALDI process, necessitating experimental balancing of their concentrations.

Polydispersity defines the spread of the components that make up a polymer. The narrowest distribution is, by definition, 1, and the closer a polymer is to this value, the more likely is it that MALDI can be used. Once the polydispersity approaches 2, it becomes difficult to obtain data and the samples should be fractionated using size exclusion chromatography and the fractions analyzed separately to obtain a correct representation of the overall makeup of the polymer.

2.2.4.1 Surface-Enhanced Laser Desorption Ionization (SELDI)

In SELDI, a variation of MALDI, special modified array surfaces are used to separate and selectively retain entire subsets of proteins directly from biological samples. The types of sample surfaces used include those that are modified chemically, e.g., with weak and strong ion exchangers, or biologically, e.g., with an antibody. Samples are first allowed to interact with the surface, and then the species that are not retained are washed away. The retained proteins are mixed with MALDI matrices and subsequently released as gaseous ions. Despite recognized shortcomings arising from lack of reproducibility and difficulties in the identification of observed species, SELDI is still used in proteomic discovery studies and for biomarker research.

2.2.5 ALTERNATIVE IONIZATION METHODS

There are numerous other ionization methods, but they have limited applications. *Fast atom bombardment* (FAB), also known as liquid secondary ion mass spectrometry (LSIMS), was one of the early methods developed for the ionization of polar molecules. FAB is based on bombarding analytes in a matrix of low volatility, such as glycerol, with accelerated energetic neutral atoms (argon or xenon) or ions (cesium) that will sputter $[M + H]^+$ ions from the surface. Although of major importance during its heyday, FAB has been superseded by ESI.

Field desorption (FD) and *field ionization* (FI) are related techniques that produce positively charged ions by using a high electric field (~10^8 V/cm) to remove an electron from the analyte using quantum tunneling. The methods impart minimal amounts of energy to the analytes, leading to the formation of molecular ions and engendering little fragmentation. FD is particularly useful for analytes with low polarity, such as long-chain alkanes. Applications include product analysis in the petroleum industry and the characterization of synthetic polymers. *Liquid introduction field desorption ionization* (LIFDI) allows the introduction of air-sensitive compounds from an inert environment, enabling the analysis of certain classes of compounds, such as organometallics, that can be hard to characterize using other methods.

2.3 MASS ANALYZERS

Once ions have been formed, either outside the mass spectrometer using API methods or within the vacuum system by EI, CI, or MALDI, the ions must be separated according to their *m/z* ratios. There are several types of *mass analyzer* with significantly different modes of operation, but all separate ions according to their *m/z* ratios, so that these ratios and their intensities can be recorded by the detector. Current mass analyzers include *quadrupole* (Q), *quadrupole ion traps* (QIT, LIT), *Fourier transformed* based (FT), *time-of-flight* (TOF), and to a much lesser extent, *magnetic field* (B).

It is now common to combine two or more analyzers within a single instrument (MS/MS) to improve and extend analytical capabilities. The combinations may involve similar analyzers (*tandem*), as in a TOF/TOF system, or be of mixed types (*hybrid*) as in a QTOF instrument.

Analyzers can also be classified based on whether or not the operating parameters must be altered to observe ions with differing *m/z* values. Mass spectrometers based on quadrupole technology—single quadrupoles (Qs), quadrupole ion traps (QITs), and linear ion trap (LITs)—as well as magnetic sector (B) systems are *scanning* instruments because the electric (or magnetic) fields of the analyzer must be varied continuously to obtain spectra. The time-of-flight (TOF) systems and Fourier transform (FT)-based ion traps (ion cyclotron resonance and orbitrap) are *nonscanning* because spectra are obtained without altering the operating parameters of the analyzer.

2.3.1 Quadrupole (Q)

Quadrupole analyzers, as their name implies, consist of a set of four rods. The rods, which are metal (molybdenum) or metal-coated ceramic with lengths of 10–20 cm and diameters up to 1 cm, are placed parallel to each other in a square with the opposite pairs connected electrically (Figure 2.21). The alignment of the rods must be precise to create the symmetrical electrical field necessary to separate ions with different m/z values. The profile of the surfaces of the rods facing the center point of the square should be hyperbolic to obtain the ideal electrical field, but cylindrical rods are also used frequently. The voltage placed on one pair of rods is comprised of a positive direct current (dc) combined with a superimposed radio frequency (rf) voltage. The other pair of rods carries a negative dc voltage with an rf component that is 180° out of phase with that on the first pair. Note that although the rods are designated as positive or negative, this a misconception because the overlaid rf voltage results in the rods constantly oscillating between positive and negative polarities.

Mass separation (also called mass filtering) is based on the fact that ions begin to oscillate upon entering the field produced by the superimposed rf and dc voltages. For any field, derived from the combination of the voltages, only ions with one specific m/z value have a stable oscillatory trajectory (*bounded oscillation*) along the axis of the quadrupole to the detector. All other ions with different m/z values develop unstable oscillation patterns (*unbounded oscillation*) perpendicular to the flight path and are lost by collision with, and discharge onto, the rods (or the prefilters, if present). Changing the dc and rf voltages progressively (the scan function),

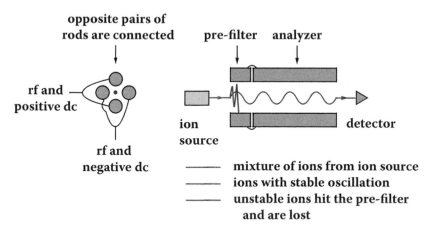

For any given set of rf and dc voltages on the opposing pairs of rods, only ions of one m/z ratio display a stable oscillation enabling them to reach the detector. Unstable ions hit the initial part of the analyzer, often a pre-filter, are discharged and lost. The pre-filter, which is connected electrically to the analyzer, is not essential but can be removed conveniently for cleaning.

FIGURE 2.21   Quadrupole (Q) analyzer.

while keeping their ratio constant, enables the scanning of a mass range yielding spectra comprised of the different *m/z* values.

The discharged ions are deposited on surfaces close to the beginning of the rods and build up over time, eventually compromising both resolution and sensitivity by distorting the field established by the rf and dc voltages. Installing a *prefilter*, which is a short section on each rod that can be readily removed for cleaning, reduces the contamination problem. The prefilters are electrically connected to the rods, are part of the analyzer, and also act as the initial point of separation and discharge of ions with unbounded oscillation.

Quadrupole analyzers have *unit resolution* throughout their mass range; i.e., one mass is discriminated from the next, be that 28 from 29 or 2,000 from 2,001. This classifies quadrupoles as low-resolution instruments which are poorly suited to obtain accurate mass data. A consequence of the poor resolution is that two ions with the same nominal mass but with different empirical formulae, and therefore different exact masses (*isobaric ions*), will not be separated, but instead, their masses will be averaged into the *m/z* observed value. Furthermore, the peaks are broad (~1 *m/z* wide), and thus it is difficult to obtain precise measurements of the centroids. The mass range of quadrupoles extends to 4 kDa for singly charged ions. These analyzers are compatible with all ionization methods except MALDI (because of the mass range limitation).

Being a scanning instrument has consequences for the modes of operation as well as for sensitivity. Consider obtaining a mass spectrum over a mass range of 1,000 Da in 1 s. Given that the scan function of quadrupoles is linear, and that only one combination of rf and dc values will allow the passage of a specific *m/z*, the time during which each *m/z* can be detected is 1 s/kDa, or 1 ms/Da. Thus, for a 1,000 Da/s scan the *duty cycle* of the instrument for each mass is 0.1% (1 ms = 0.1% of a second); this means that a great deal of time is sacrificed to obtain a spectrum.

There is, however, an important advantage of the available duty cycle when Q instruments are used for quantification of compounds whose mass spectra are known. Once an *m/z* value characteristic of an analyte is obtained, the rest of the spectrum becomes irrelevant. The voltages on the quadrupole rods can then be set at values to enable the passage of only the *m/z* that is specific to the analyte. The paths of all other ions will be unstable and those ions will be lost; i.e., the quadrupole is acting as a mass *filter* allowing passage of only the selected ions to the detector. Under these conditions the duty cycle for the selected ion is 100% and the analyzer is at its most sensitive for the chosen *m/z* value. This mode of operation is called *selected ion monitoring* (SIM). In essence, the quadrupole can be set to scan only one mass unit, something that cannot be done with TOF analyzers where all masses must traverse the analyzer even if they are irrelevant to the task at hand (Section 2.3.3). Duty cycles will be revisited with respect to data collection using TOF and FT analyzers.

The fact that the fields in Q analyzers are electronically derived and can be manipulated rapidly and selectively has several advantages: (1) fast scanning (e.g., 0.1 s/mass decade, or 10 Hz for an *m/z* 50–500 scan), (2) essentially no reset time between scans, (3) easy specification of mass ranges, and (4) ability to monitor one or several selected masses anywhere within the mass range with rapid jumping between different *m/z* values. These advantages are utilized fully in GC-MS and LC-MS both for qualitative analyses and for high sensitivity quantification using

SIM. Additional benefits of Q analyzers include ease of use, small footprint, and lower cost. Disadvantages include the low duty cycle (per m/z) and the lack of accurate mass capability.

2.3.1.1 Quadrupoles as Ion Guides (q)

When a quadrupole is used as a mass analyzer, as described above, it is operating in a *narrow band* pass mode, filtering one m/z value from another. Alternatively, the same rf voltages can be applied to the quadrupole, but without being combined with a dc field; this is the *wide band* pass mode. Under these conditions all ions, e.g., those emerging from the ion source or from Q1 in a triple quadrupole (Section 3.3.3.1) that enter the q are transmitted, regardless of their m/z values. Hence, here the q acts as an *ion guide* (rather than an analyzer) delivering ions between two points. (The lowercase q designates that the quadrupole is used as an ion guide rather than a mass analyzer.) When quadrupole are used as ion guides, it is common to see different combinations of rods, such as hexapoles or sometimes octapoles, because ion transmission is more efficient with these formats.

The versatility of quadrupoles in either the Q or q modes is of particular relevance in multianalyzer instruments. The capability of switching rapidly between the Q or q modes of operation, as well as being able to use Q in scanning or static format, enables experiments in which different sets of ions are passed from one analyzer to a second, depending on how the first analyzer is used. Some strategies involving these features are described in Section 2.3.7.2.1.

2.3.2 ION TRAPS

2.3.2.1 Quadrupole Ion Trap (QIT)

Although both quadrupoles and *quadrupole ion traps* are based on the stability of ions in the fields created by the combination of rf and dc voltages, the functionality of the two systems is opposite. In the Q analyzer only ions stable in a specific field are passed to the detector, while the unstable ions are lost. In the QIT analyzer all ions are trapped in a stable trajectory within the cell. Selected m/z values are detected as they are rendered unstable by an applied voltage and are eventually ejected from the trap onto the detector. An obvious advantage of the QIT is that the trapped ions can be maintained inside the cell for extended periods and are therefore available for additional experimentation.

The QIT is a three-dimensional quadrupole comprised of a toroidal ring electrode and two end caps, all machined to have hyperbolic inner surfaces (Figure 2.22). Ions are produced in an external ion source and only a controlled number is allowed to enter the trap through one of the end caps (by monitoring the ion current). A three-dimensional quadrupole field, created by a combined rf/dc field on the ring electrode and the grounded end caps, restrains all injected ions in a complex sinusoidal path within the cell. When the QIT is operated with helium in the trap (pressure ~1 mTorr) the gas causes the ions to cluster and remain close to the center of the cell where the electrical field is most symmetrical, resulting in the ions staying in their most stable orbits. Collisional cooling of the ions by the helium also assists in maintaining

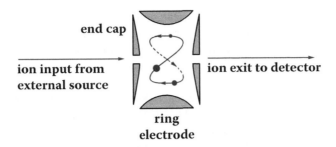

Ions injected into the trap are held by rf and dc voltages placed on the hyperbolic ring electrode and end caps.

Ions move in a complex sinusoidal path where there are crossover points. Space-charge effects occur at the crossover points of the ion paths when too many ions are present.

By changing the voltages progressively, ions with different m/z values are ejected sequentially onto the detector, generating spectra.

CID fragmentation takes place when a specific m/z is isolated in the trap and excited in the presence of a neutral gas, providing MS/MS data.

The isolation/fragmentation cycle can be repeated to obtain MS^n data.

FIGURE 2.22   Quadrupole ion trap (QIT) analyzer.

stable ion paths. Changing the amplitude of the rf on the ring electrode increases the radius of the ion paths within the cell. The combination of an amplified rf voltage on the ring electrode with an rf voltage on the end caps ejects ions from the cell through the end cap beyond which the detector is located. The combination of the rf voltages necessary to render an ion unstable is specific to each m/z. The technique, called *resonance ejection*, is used to obtain mass spectra by ejecting ions sequentially, according to their m/z values. It is also possible to retain a specific band of m/z values by ejecting all ions above and below the selected range.

Alternatively, ions with a specific m/z may be selected, and maintained in the trap with a stable trajectory, while all others with different m/z values are removed. The retained ions can be fragmented using collision-induced dissociation by introducing an additional pulse of gas into the trap. A scan of the product ions provides an MS/MS spectrum. This isolation and fragmentation cycle can be repeated to provide MS^n spectra. The method is usually pursued only as far as MS^3, although up to MS^6 and higher are possible, because there is a significant reduction in the signal as the number of iterations increases. For instance, if there are 10^3 ions of a given m/z in scan 1 (MS) and 10 fragments each with 10^2 ions in scan 2 (MS/MS), then there are only 10^2 ions available at a specific m/z for scan 3 (MS^3), a 10-fold reduction in sensitivity.

The ions within a QIT follow complex sinusoidal paths, thus there are multiple locations where ion paths cross (Figure 2.22). The density of the ions at a crossover location can reach a level where the ion current distorts the electrical field within the cell, causing space-charge effects that degrade the quality of the data. Consequently, there is a limit to the number of ions that can be held in a QIT, and this affects the sensitivity and dynamic range of these analyzers. The actual number of ions that can be held in the trap varies, depending on whether the intention is to store ions, isolate specific m/z, or obtain spectra. The last of these is the most stringent, with approximately 2,000 ions allowed.

QITs are remarkable instruments because spectra can be obtained, ions stored, and sophisticated MS/MS experiments conducted, all in a cost-effective way (with a small footprint). One limitation, applicable to all ion traps, is the so-called *one-third rule*, which states that in CID MS/MS it is not possible to detect product ions that are less than about one-third of the value of the precursor ion; e.g., for a precursor ion of m/z 900 the detection limit for product ions is m/z ~300. This limit can be frustrating; e.g., it is not possible to detect the immonium ions formed from amino acids during CID of a peptide, nor can QIT be used with CID for certain techniques, such as iTRAQ (Section 3.5.19.3). Similarly to quadrupoles, from which they are derived, QITs have limited resolution. Alternative fragmentation induced by rf pulsing provides a type of scan in which the one-third rule does not apply and low mass ions can be observed.

The physical format of the QIT is different from that of other tandem instruments because the two mass separation steps needed to obtain MS/MS data are *sequenced in time* and use the same analyzer. There are other instrument formats for the acquisition of MS/MS data, but in those systems the analyzers are placed sequentially, i.e., MS/MS *in space* (Section 2.3.7).

2.3.2.2   Linear Ion Trap (LIT)

The limit on the number of ions that can be placed in a QIT before space-charge effects distort the spectra and reduce sensitivity led to the development of the linear ion trap (LIT). In the LIT ions oscillate in a linear fashion along the length of a quadrupole, removing many crossover points where the major space-charge problems occur in a QIT (Figure 2.23). Grids at each end of a quadrupole, with voltages that can be manipulated, are used to allow ions to enter the LIT, and to reflect and maintain the ions in the cell, and then enable their exit either onto a detector or into a second analyzer. Alteration of the voltages on the rods of the quadrupole enables radial ejection of ions with unwanted m/z and the convenient selection of a specific ion that can then be quantified, fragmented (MS^n), or passed on to a second analyzer. These capabilities are similar to those of the QIT, but the LIT allows the utilization of a much larger number of ions, thus providing significant improvement of detection limits (approximately 30,000 ions when collecting spectra). However, the limitations in resolution, mass accuracy, and the one-third rule remain.

2.3.3   Time-of-Flight (TOF) Analyzer

Although modern TOF analyzers may be rather complicated, the basic principle is simple: they are just what they say they are. Pulses of ions are accelerated from the source into the analyzer tube, and the time for an ion to travel through a field-free region to the detector is measured. The time-of-flight for the passage of an ion in a TOF analyzer is a function of its momentum, and therefore its m/z. The acceleration voltage and, consequently the kinetic energy (momentum), is the same for all ions,

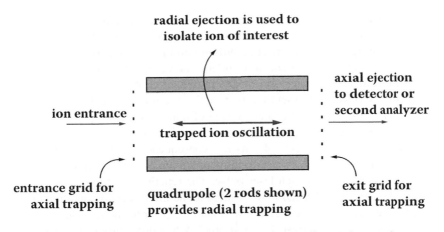

**radial ejection is used to
isolate ion of interest**

**axial ejection
to detector or
second analyzer**

ion entrance

trapped ion oscillation

**entrance grid for
axial trapping**

**quadrupole (2 rods shown)
provides radial trapping**

**exit grid for
axial trapping**

Selective use of the rf and dc voltages on the rods can be used
to axially eject ions in m/z sequence to obtain a spectrum, to
pass ions onto another analyzer, or to isolate a specific ion by
radially ejecting all other ions with different m/z values.
Voltages on the entrance and exit grids enable the accumulation
and maintenance of ions inside the trap.
LIT are more sensitive than quadrupole ion traps (QIT) because
there are fewer crossover points in the ion paths so the LIT can hold
more ions before space-charge effects disrupt instrument
performance.

FIGURE 2.23 Linear ion trap (LIT) analyzer.

thus those with the lowest m/z will travel fastest and arrive at the detector first, followed by the sequential arrival of ions with successively higher m/z.

TOF analyzers have been around since the 1950s but fell into disuse because of their poor resolution (limited to ~200, i.e., separation of m/z 200 from m/z 201). The main reason for the inadequate resolution was the broad spatial and kinetic energy distributions of the ion packets produced by the laser-based ion sources. The momentum of an ion, as it enters a TOF analyzer, is a function of the energy it receives during ionization, its location at the time that it is accelerated, and the accelerating voltage. In early instruments it was not possible to line up the ions in the same plane prior to their acceleration into the analyzer. Furthermore, when molecules with the same empirical formula are ionized, it is inevitable that the ions will have a range of energies and that these differing energies will cause the ions to travel at different speeds and arrive at the detector at different times, also resulting in a loss of resolution. Early TOF instruments were further hampered by the lack of high-speed electronics and computing. TOF analyzers underwent a resurgence in the 1990s with the development of means to account for the spatial and energy differences of ions using delayed extraction and the reflectron, as well as improvements in electronics and computing.

TOF analyzers are nonscanning analyzers because the operating parameters of the analyzer need not be altered to obtain spectra. An important consequence of being nonscanning is that ions must be pulsed into the analyzer as discrete packets that contain all *m/z* values produced in the ion source. In addition, all ions in each packet must clear the analyzer prior to the introduction of the next packet of ions. The time for the passage of a set of ions is on the microsecond scale, and there will be distortion of the flight paths and overlap of data if a second packet of ions is allowed to enter the analyzer before the first has arrived at the detector.

The need to generate pulses of ions has consequences, depending on whether the ion source produces pulses or a continuous stream of ions. In MALDI sources, ions leave in packets upon each laser pulse. The repetition of the pulses is on the millisecond scale, thus MALDI is well suited to the microsecond operation of TOF analyzers. Each set of ions will clear the flight tube prior to the introduction of the next packet of ions. In contrast, ESI and other API sources produce ions continuously. The need to form ion packets has been accommodated by the development of *orthogonal* TOF analyzers (discussed below).

Delayed extraction (also called *timed ion delay*) in MALDI-TOF instruments addresses the spatial distribution problem by aligning and concentrating ions at an *electronic gate* prior to their release into the analyzer. Another benefit of delayed extraction is that there is *collisional cooling* of the ions during the lining up process that reduces their energy distribution. The collisions transfer energy from the ions to residual molecules in the vapor phase, thereby reducing the vibrational energy added during ionization. The collisions are low-energy interactions that limit fragmentation.

In a simple, linear MALDI-TOF instrument the ions move in a straight line from the source to the detector (Figure 2.24). The illustration shows the scattering of ions as they come off the surface of the sample plate during the MALDI process. The ions are then "lined up" at the delayed extraction grid prior to their release into the analyzer tube. The time of the delay is in the 50–750 ns range, depending on sample type, with larger molecules requiring longer delay times. Once lined up at the grid, all ions are subjected to the same acceleration voltage and have the same distance to travel to the detector; consequently, they are separated as a function of their *m/z* values ($z = 1$).

The use of the delayed extraction grid improves resolution by reducing the spatial differences among the ions and, to some extent, the inhomogeneity of their energy content (Figure 2.24). However, despite the collisional cooling at the delayed extraction grid, there will still be differences in the distribution of the acquired energies for any given set of ions of the same empirical formula. As a result, ions with the same *m/z* but slightly different energies will travel at different speeds and will spread out in the flight tube, causing differences in their arrival times at the detector, and thus a range (albeit small) of measured masses. In other words, the resolution of the linear TOF instrument is limited by the spreading out of ions that have the same empirical formula and *m/z* value.

Reflectrons (ion mirrors) are used to increase the resolution, as well as the ability to obtain accurate mass measurements, by counteracting the effect of the energy spread not compensated for by collisional cooling at the delayed extraction grid (in MALDI sources). A reflectron is a cylinder composed of concentric rings and grids

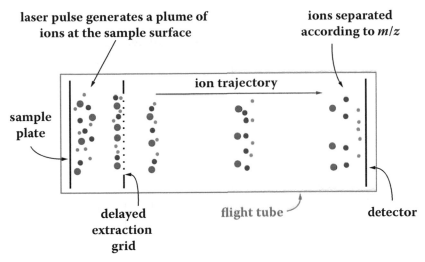

laser pulse generates a plume of
ions at the sample surface

ions separated
according to *m/z*

ion trajectory

sample
plate

delayed
extraction
grid

flight tube

detector

Ions leaving the sample plate are lined up at the delayed
extraction grid that reduces both their spatial and
energy spreads, the latter through collisional cooling,
before their release into the flight tube.

The momenta of all ions (which are singly charged) are
the same as they enter the flight tube. Lighter ions will
travel faster so the arrival of ions at the detector will be
in the order of their *m/z* values.

FIGURE 2.24 Behavior of ions in a linear MALDI time-of-flight analyzer.

onto which a voltage gradient is applied, thereby creating an electrical field of increasing strength into which the ions enter. The field is set to the same polarity as that of the ions analyzed. Ions entering the electrical field are slowed (because like charges repel each other), their direction is reversed, and eventually they are repelled back, out of the reflectron. The principle of the process is that if two ions of the same *m/z* but with slightly different energies are made to reverse direction in an electric field, then the more energetic ion will require an electric field of slightly higher strength to cause the reversal. The more energetic ion will therefore take a longer path through the reflectron and will exit the reflectron at a slightly higher speed than will the less energetic ion. Accordingly, there will be a focal plane where the two ions arrive at the same time, thus accounting for the differences in their initial energies (Figure 2.25).

Delayed extraction and a reflectron each provide about 10-fold increases in the resolution of MALDI-TOF instruments. Delayed extraction improves resolution from ~200 to ~2,000, and incorporation of a reflectron increases resolution from ~2,000 to ~20,000. As resolution increases, so does the ability to accurately calibrate the mass scale. In linear mode (i.e., without a reflectron), only nominal mass measurements can be obtained. Accurate mass determinations become possible once a reflectron is incorporated.

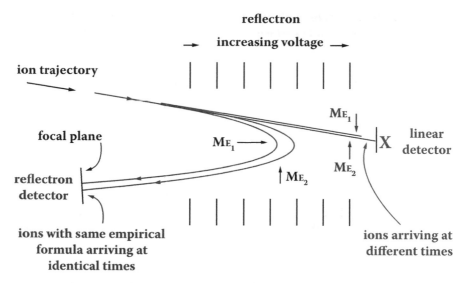

ME$_1$ and ME$_2$ are two ions with the same molecular structure but with
different energies (E$_1$, E$_2$). These ions will travel at different speeds
and arrive at a linear detector (at point X) at different times, despite
having left the ion source at the same time. ME$_2$ is shown arriving
at X before ME$_1$ so these ions will appear to have different masses
(ME$_2$ will be lighter than ME$_1$).
When a reflectron is incorporated, the energy spread of structurally
identical ions is compensated for by the depth that the ions travel
into the reflectron prior to reversing direction and arriving
simultaneously at the reflectron detector, placed at the focal plane.

FIGURE 2.25   Principle of the operation of a reflectron.

There is an important caveat for reflectrons: they are effective only to ~5 kDa,
after which the analyzers must be used in linear mode with a subsequent loss in
resolution. The loss of resolution is most relevant when MALDI sources are used
because the method produces only singly charged ions, and thus the advantages of
the beneficial properties of the reflectron are lost in the analysis of proteins. On the
other hand, the multiply charged ions formed during ESI are usually <*m/z* 5,000, so
they can be focused with a reflectron.

As mentioned earlier, the use of TOF is more complicated with ESI and other
API sources, than with MALDI, because these methods produce ions continuously;
thus, packets of ions must be excised from the ion stream and then passed through
the flight tube prior to the introduction of the next packet of ions. To overcome this
problem, the analyzer is set at a right angle (orthogonal) to the ion beam emerging
from the source. Ion packets are created in a *pusher* region using a pulsed voltage.
The trajectories of the ions entering the analyzer have both horizontal and vertical
components. The former arise from the direction that ions are traveling from the

source, and the latter from the pusher voltage acting at 90° (Figure 2.26). A compromise has to be made when there is orthogonal injection from a continuous stream of ions. Because each packet of ions must completely clear the analyzer before another one can be introduced, some of the stream of ions from the source will inevitably pass through the pusher region without being sampled. As only part of the continuous beam can be analyzed, this limits the duty cycle, and therefore the sensitivity, of orthogonal TOF analyzers.

Both the starting points and energies of the ions must be controlled as precisely as possible to attain the highest resolution. The focusing and control of the ion beam in the pusher region defines the starting point of the analyzer when using continuous ion generation (API, EI, CI). Positioning the ions in the pusher region prior to their injection into the analyzer is equivalent to the process occurring at the delayed extraction grid in MALDI sources. The ion packets excised by the pusher are accelerated orthogonally into the flight tube where ions of different m/z values are separated. To obtain the energy focusing needed to improve resolution, the ion path must include passage through a reflectron. The combination of the horizontal and vertical motion components of the path of the injected ions and the use of a reflection results in a V-shaped ion path that is characteristic of ESI-TOF instruments (Figure 2.27).

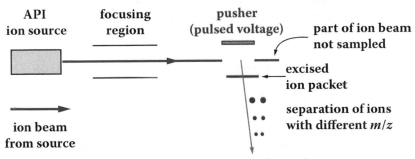

API ion source   **focusing region**   **pusher (pulsed voltage)**   **part of ion beam not sampled**

excised ion packet

ion beam from source

separation of ions with different m/z

direction of ions in the analyzer is composed of horizontal and vertical components imparted by the source and pusher voltages.

Ions are generated continuously in API sources, thus, the individual packets of ions required in TOF analysis must be excised from the ion beam and accelerated orthogonally into the analyzer using a pulsed pusher voltage.

Only part of the continuous ion beam can be sampled, as the excised ion packet must traverse the analyzer before another set of ions is introduced. The duty cycle of the orthogonal injection process limits the sensitivity of orthogonal TOF instruments.

FIGURE 2.26   Injection of ions into an orthogonal TOF analyzer.

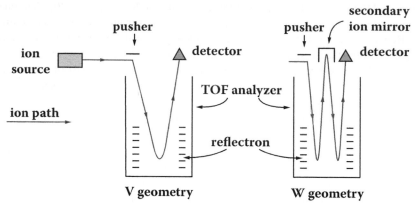

Ions arriving from the source are accelerated orthogonally into the analyzer by a pulsed voltage from the pusher.

Ions separate based on their momenta. As they travel through the analyzer the lightest, fastest ions (lowest m/z) will arrive at the detector first.

Some horizontal momentum imparted in the source remains so that ions travel at an angle into and out of the reflectron thus attaining a characteristic 'V' trajectory.

Increasing the distance that the ions travel improves mass resolution, e.g., by using 'W' geometry.

FIGURE 2.27   Principles of orthogonal TOF analyzers with V and W geometries.

Increasing the length of the analyzer flight path should be an obvious way to increase resolution, given that the separation of ions with differing m/z values, i.e., the resolving power of the analyzer, is a function of the arrival time of ions at the detector. Indeed, there are instruments (with both MALDI and ESI sources) with flight paths several meters (even 20 m) long. These extended flight path systems would be large and inconvenient if the ion path were of the V type. Accordingly, an increase in path length is achieved by using an additional reflectron to give the ion path a W geometry (Figure 2.27) and more complex, multipass formats. A compromise associated with increasing the flight path is that the further ions have to travel, the greater are the chances that some will be lost and the sensitivity reduced.

TOF analyzers have several important advantages that are making them a mainstay of modern mass spectrometry, including:

1. Theoretically unlimited mass range, currently ~350 kDa; this is particularly important with MALDI sources where ions are singly charged.
2. Each packet of ions is analyzed independently, on a μs scale, resulting in the collection of tens of thousands of spectra per second. It is not practical to observe such spectra individually, but they can be grouped and displayed by the data system over various time spans, permitting specialized applications, such as studies of reaction kinetics of explosives on a 50 ms timescale.

3. Resolution is independent of the time period over which data are collected.

4. TOF analyzers are particularly useful when interfaced with chromatography where peak widths are measured in seconds; it is important to obtain as much MS and MS/MS data as possible during that time period to fully characterize each individual chromatographic peak. In addition, the rapid acquisition of spectra at high resolution enables the collection of all data in accurate mass measurement mode that can be further interrogated, e.g., to assist in the deconvolution of the components of overlapping LC peaks.

2.3.4 FOURIER TRANSFORM ION TRAPS

At the top end of the resolution scale are the Fourier transform (FT) instruments, including the orbitrap and the ion cyclotron resonance (ICR) analyzers. Both these analyzers are ion traps; however, unlike the quadrupole-based traps, and indeed any other class of analyzer, the analytical cell in FT instruments is also the detector, with the ions being recorded as an *image current* (also called a *transient*). The generation and interpretation of the image current is discussed together with the other types of ion detectors (Section 2.4.3). The orbitrap and ICR are described below in order of increasing resolution. The orbitrap is a single analyzer, although the name is commonly applied to a configuration that also includes a linear ion trap, making it a hybrid system (Section 2.3.7.2.2).

2.3.4.1 Orbitrap

The most recent commercial mass analyzer, the orbitrap, is an electrostatic ion trap comprised of a spindle-shaped inner electrode and a split outer electrode (Figure 2.28). Ions, typically from ESI or APCI, are collected in a specialized component called the *C-trap* and then injected into the orbitrap as high-speed pulses. The speed provides the ions with the centrifugal force necessary to counteract their attraction to the central electrode. Upon entering the cell, the ions follow a complex trajectory composed of two components: a rotary motion around the inner electrode and an axial oscillation along the same electrode that is at right angles to the rotary component. It is the oscillatory motion that induces a current in the split outer electrode. Because the periodicity of the axial oscillation is proportional to the *m/z* of the component ions, an image current is generated, recorded, and eventually interpreted for its harmonic composition using FT analysis (Section 2.4.3).

The resolution of orbitraps can exceed 200,000. This is less than that of FT-ICR systems, but is still much higher than any other analyzer. Other advantages include mass measurement with high accuracy and a dynamic range of $>10^3$. A disadvantage is that the number of ions that can be held in the trap is limited, and this may affect the detection of minor sample components, particularly when there are multiply charged analytes, as all the isotope ions within each of the charge states contribute to the total number of ions. Orbitraps are much less expensive than FT-ICR systems and are easier to maintain, because there is no need for a superconducting magnet and its associated cryogens.

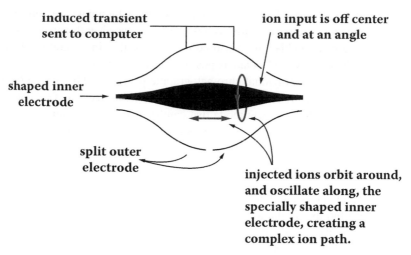

induced transient
sent to computer

ion input is off center
and at an angle

shaped inner
electrode

split outer
electrode

injected ions orbit around,
and oscillate along, the
specially shaped inner
electrode, creating a
complex ion path.

**Ions are injected at an angle and offset from the center of the trap.
The momentum of the ions causes them to orbit around, and
oscillate along, a central spindle-like electrode.
The lateral oscillation of the ions along the inner electrode induces
a transient (image) current in the split outer electrode.
The recorded image current is interpreted using Fourier transform
analysis to provide *m*/*z* values and intensities.**

FIGURE 2.28   Orbitrap analyzer.

2.3.4.2   Fourier Transform–Ion Cyclotron Resonance (FT-ICR)

The FT-ICR analyzer is based on the well-known observation that ions move in cir-cular paths in a magnetic field. The frequency of the rotation depends on both the m/z of the ions and on the strength of the magnetic field. The direction of the motion can be visualized using Fleming's *left-hand rule*, which states that a positive ion traveling through a magnetic field, set at 90° to the ion's trajectory, will be curved in a clockwise direction (Figure 2.29). The radius of the curve is proportional to the mass of the ion.

The analytical cell is located within the core of a super-cooled, superconducting magnet. Commercial instruments have magnets in the 7–15 Tesla range. Ions from the source, or from another mass analyzer, move to the cell (trap) by passing through a series of pumping stages of increasingly high vacuum. The ions are trapped on enter-ing the cell and move in an orbital path, possibly for hours, in an ultra-high vacuum (10^{-10} Torr). Trapping is accomplished by applying small voltages at the entrance and exit of the analytical cell. The ions rotate in a cell that is divided into quadrants with opposing pairs connected electrically (Figure 2.30). In one pair of plates (*receiver* plates) the charge of the cycling ions induces a current. The second pair of plates (*transmitter* plates) is set so that the applied voltages are 180° out of phase. While the ions are cycling close to the center of the cell they are too far away from the receiver

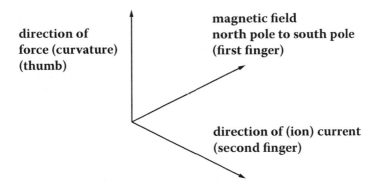

**direction of
force (curvature)
(thumb)**

**magnetic field
north pole to south pole
(first finger)**

**direction of (ion) current
(second finger)**

**If the thumb, first, and second fingers of the left hand are held
at right angles to each other, the thumb shows the direction
of curvature of a positive ion in a magnetic field.**

FIGURE 2.29   Fleming's left-hand rule.

plates to induce a current. The ions must be sorted by m/z and subsequently moved increasingly closer to the receiver plates prior to detection. This is accomplished using a ramped rf voltage pulse sent via the transmitter plates. Energy is absorbed by ions of a given m/z only at a particular ratio of the rf pulse and magnetic field. Changing the ratio enables bringing together (coalescing) the ions at each m/z into individual packets. The frequency of the rotation of each set of ions is related to its m/z, while the number of ions in each packet determines the amplitude of the induced signal. The ions spiral back to their stable trajectory as the orbits decay after the pulse. The decay is reflected in the characteristic shape of the recorded image current. Mass spectra are obtained, as with the orbitrap, by deconvoluting the image current using FT analysis.

The unique advantages of FT-ICR analyzers are ultra-high resolution, up to 3 million, and the ability to determine masses to a high degree of accuracy (fractions of a mDa), higher than with any other type of mass analyzer. Other significant advantages include that both ESI and (to a degree) MALDI sources may be used, all masses are detected simultaneously, mass fragmentation studies can be carried out up to MS^4, and operational modes can be changed easily, e.g., from low to high resolution, as only electrical parameters need to be adjusted.

FT-ICRMS is often the best (sometimes only) solution for certain studies. Examples include top-down proteomics where intact, undigested proteins are analyzed (Section 3.5.1.3) and the analysis of highly complex mixtures of small molecules, such as crude oil, where chromatography is of no value because of the presence of too many related components with similar retention times (Section 4.6).

A difficulty with both the orbitrap and ICRMS is that their ability to provide very high resolution depends on keeping the ions in the analyzer for considerable periods of time, and this may not be compatible with the chromatographic timescale. For instance, when analyzing an ion of m/z 800 with a 9.8 Tesla magnet, a 5 s transient is required to attain a resolution of ~500,000, while a 0.5 s transient provides a

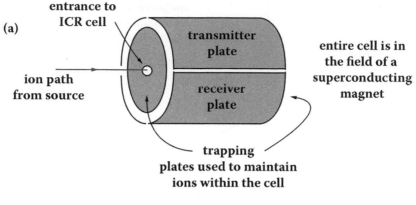

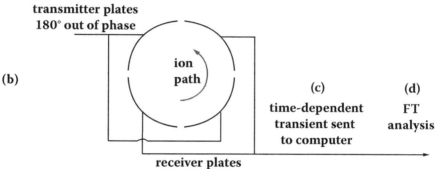

(a)   The ICR cell is located within a superconducting magnet with the magnetic field perpendicular to the cell. Ions are injected into the cell at one end and are trapped by plates (grids) at both ends.

(b)   Ions travel in a circular path. Their proximity to the receiver plates is controlled by pulses from transmitter plates.

(c)   The time-dependent transients induced in the receiver plates have component frequencies that are proportional to the m/z with amplitudes that reflect ion intensities.

(d)   Fourier transform analysis of the recorded transients provides the mass spectra.

FIGURE 2.30   (FT) ion cyclotron resonance (ICR) analyzer.

resolution of only 47,000 (also available with high-end TOF systems). Other disadvantages include the need for superconducting magnets (in ICR systems) and very high-performance vacuum systems, making these instruments very expensive, up to US $2 million. Maintenance of the magnets in ICR instruments is costly, as the magnet must be kept appropriately filled with liquid helium and liquid nitrogen. Loss of the cryogens results in quenching of the magnet, a very expensive accident.

2.3.5 Ion Mobility Analyzers

Mass spectrometrists have always been concerned with the measurement of the mass and intensity of analyte ions. Investigation/utilization of the *shapes* of molecules is now possible with *ion mobility* techniques that utilize differences in the cross sections of ions as they move through a gas. Think in terms of two pieces of paper, one crumpled and the other flat. If dropped at the same time, the crumpled one will hit the floor first because it will encounter less air resistance than the flat piece. A similar situation applies to ions with different shapes as they travel through a gas. Although ion mobility has been examined with home-built instruments for years, only recently has this type of analyzer become available commercially. There are two significantly different types, the *high-field asymmetric waveform ion mobility spectrometer* (FAIMS) and the *ion mobility separator* (IMS). The FAIMS separator is placed between the ion source and the analyzer, while the IMS cell is located between the analyzers of an MS/MS instrument.

2.3.5.1 High-Field Asymmetric Waveform Ion Mobility Spectrometer (FAIMS)

In FAIMS, voltage waves are placed on two electrodes, at atmospheric pressure, using a high-voltage (kV) asymmetric waveform at radio frequency. Figure 2.31a shows both the asymmetry of the applied voltages and also that the areas under the positive and

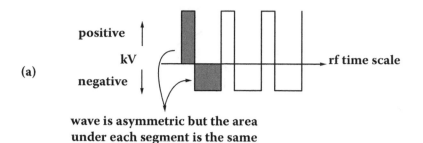

(a)

wave is asymmetric but the area
under each segment is the same

The asymetric voltage wave (above) is applied to
the parallel analyzer electrodes.

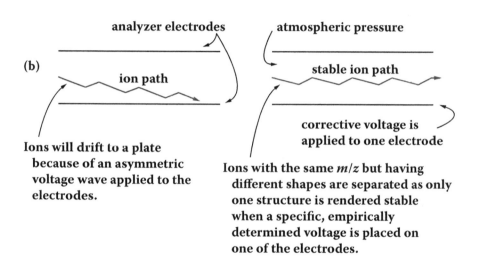

(b)

Ions will drift to a plate
because of an asymmetric
voltage wave applied to the
electrodes.

Ions with the same *m/z* but having
different shapes are separated as only
one structure is rendered stable
when a specific, empirically
determined voltage is placed on
one of the electrodes.

FIGURE 2.31   High-field asymmetric waveform ion mobility spectrometer (FAIMS) analyzer.

negative sections of the wave (i.e., applied voltage × pulse time) are the same, although
their times and magnitudes are different. When injected into such a field, ions will not
only be compressed and focused, but will also drift to one or the other of the electrodes
(depending on their shape and structure) where they will be discharged and lost. When
a *correction voltage* (dc) is applied to one electrode ions of a given structure will have
a stable trajectory emerge from the analyzer (Figure 2.31b). These ions are then introduced into a separate, conventional mass analyzer. The correction voltage is structure
dependent and must be determined empirically.

The ability of FAIMS to filter ions according to their shape is utilized to improve
the signal-to-noise ratio when determining the presence, or to obtain the concentration,
of a compound in a complex matrix, e.g., analyte y. When the *m/z* of y is monitored, the
signal observed will be from both y and other interfering compounds with the same
m/z. The various isobaric ions (even isomers) have different shapes (cross sections),
and therefore different resistances as they move through the gas in the analyzer. The

correction voltage applied to the *m/z* of y is specific providing a stable trajectory to the y ions, while any interfering compound(s) are lost on the walls of the analyzer, thus leading to an improvement in signal-to-noise ratio for the y compound.

2.3.5.2 Ion Mobility Separator (IMS)

The IMS is a completely different type of ion mobility analyzer compared to FAIMS. The IMS cell is placed inside the vacuum chamber between the two mass analyzers of an MS/MS system, operates at mTorr pressures, and consists of a series of plates along which voltage waves travel. The cell is filled with nitrogen, flowing in the direction opposite to that of the voltage waves. Entering ions are pushed along by the voltage waves against the counterflow of nitrogen (Figure 2.32). The speeds at which ions with the same *m/z* (e.g., isomers) travel through the nitrogen gas is a function of their three-dimensional shapes. The resistance created by the nitrogen flow results in a reduced ability of the voltage wave to carry ions with larger cross-sectional areas; those ions fall off the back of the wave, only to be pushed on by the next wave. Structures with different cross sections become separated as the retardation mechanism happens successively through the cell. The separation process is illustrated in Figure 2.32 by the red balls and green oblongs, with the latter being retarded because of their larger cross-sectional area. IMS can be considered a *voltage surfing* process by which more compact species ride the voltage waves more efficiently than the less compact species, and emerge from the cell more rapidly. After leaving the cell the ions enter a TOF analyzer where their *m/z* ratios and intensities are recorded.

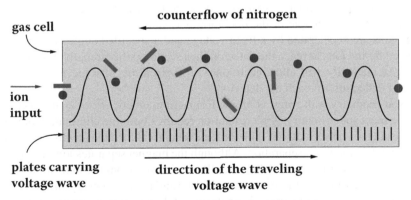

A series of voltage waves passes along a set of plates contained within a gas cell (mTorr of nitrogen). Isomeric ions with different cross sections (i.e., different shapes) will be separated: those ions with the larger cross sections are retarded by their interaction with the gas molecules while compact species (smaller cross section) encounter gas molecules less frequently and pass through the analyzer more rapidly than ions with larger cross sections.

FIGURE 2.32 Separation of ions in an ion mobility analyzer.

The time scale for mobility separations is in milliseconds; thus, IMS can be combined conveniently with LC-TOF instruments where the LC peaks are seconds wide and the TOF analyzers operate on a microsecond scale. Compounds eluting from the LC column are ionized, exit the source, and are stored in an ion trap prior to their release into the IMS. During the milliseconds that the first set of ions is traversing the IMS, a second packet of ions (from the source) is accumulated in the trapping region. At the other end of the instrument, the TOF uses orthogonal ion injection, as do other ESI-TOF systems. The microsecond timescale of the TOF enables the collection of spectra for the separated compounds emerging from the IMS.

The main advantage of ion mobility analyzers is the addition of another dimension of separation to mass spectrometry. FAIMS improves signal-to-noise ratios by removing isobaric ions that have three-dimensional shapes that are different from those of the analyte. IMS enables the measurement of the cross-sectional areas of ions in addition to the dimensions of retention time, mass and intensity provided by the chromatograph and mass spectrometer, respectively. IMS can also improve the quality of spectra by separating species that overlap chromatographically and would otherwise give mixed spectra.

2.3.6 MAGNETIC SECTOR ANALYZERS

Magnetic sector analyzers hold a fabled historical position, as they were the first type of mass analyzer developed. Once the mainstay of mass spectrometry, these analyzers are now consigned to specific applications, e.g., the analysis of the chlorodibenzodioxins and related species. As with FT-ICRMS (Section 2.3.4.2 and Figure 2.29), the principle of the operation is covered by Fleming's left-hand rule. When ions are accelerated by a fixed voltage and a specified angle of movement is required for the ions to reach the detector, then scanning the strength of the magnetic field systematically will allow each ion with a given m/z to achieve the necessary angle of deflection to reach the detector.

Fixed geometry is the most common format of magnetic sector analyzers used in organic mass spectrometry. High resolution requires both the elimination of the differences in the energies of ions with the same empirical formula (compare with TOF, Section 2.3.3) and mass separation. Although the magnet separates ions according to their m/z, energy differences (albeit small), originating from the ionization process, remain between ions of the same empirical formula. These energy differences can be eliminated with an electrostatic analyzer (ESA) where more energetic ions have to travel a longer path prior to reaching a focal point (Figure 2.33). Depending on the geometry of the analyzer, an ESA can be placed before or after the magnet, or one can be placed in each location. The process in an ESA is similar to the operation of reflectrons in TOF systems (Section 2.3.3, Figure 2.25).

Combinations of magnetic and electrostatic analyzers have been used to make instruments with high resolution capable (with appropriate calibration) of accurate mass measurements. Because they are scanning instruments, magnetic sector analyzers suffer from duty cycle limitations, as do quadrupole analyzers. Furthermore, because the scan speed of the magnetic field is proportional to resolution, additional time is required to obtain a spectrum as resolution is increased. Scanning the

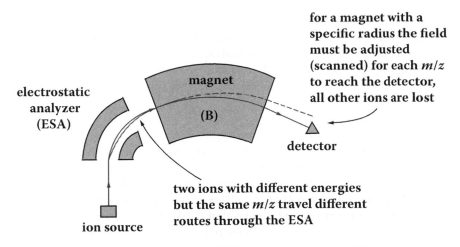

The magnetic field required to deflect an ion through a defined angle
is proportional to the momentum (therefore the mass) of the ion.
The ESA reduces the energy spread of ions with the same empirical
formula because the more energetic ions travel a longer path to a
focal point that is prior to the magnet. The ESA may also be placed
after the magnet in which case the focal point is before the detector.
The combination of energy and magnetic analyzers (double-focusing)
enables high resolution and, with appropriate calibration, accurate
mass measurement.

FIGURE 2.33 Double-focusing high-resolution magnetic sector mass spectrometer.

accelerating voltage can be used for accurate mass measurements, but this strategy is
applicable only over a limited mass range because of sensitivity and focusing problems encountered as the voltage is altered.

2.3.7 MULTIANALYZER (MS/MS) SYSTEMS

While the development of ESI and MALDI provided techniques that ionize polar
compounds effectively the spectra provide little or no structural information because
molecular species are the only ions formed. The need to investigate not only the mass
but also the *structure* of ions was an important factor in the development of multianalyzer mass spectrometers. There are several formats of *MS/MS systems* based on
combinations of quadrupole, ion trap (Q and FT), and TOF analyzers. Combinations
that use the same type of analyzers are *tandem* instruments, whereas different separation methods are used in sequence in *hybrid* systems. In MS/MS instruments, the
first analyzer is used (usually) to select an ion of interest that is then passed, with
or without fragmentation, into a second analyzer, which is often of higher resolving
power and can be used for accurate mass measurement.

2.3.7.1   Tandem MS/MS Instruments

Tandem instruments include the triple quadrupole (QqQ), QIT, LIT, and TOF/TOF systems. In the QqQ and TOF/TOF the analyzers are separated *in space*; these combinations are described below. In QIT and LIT the same analyzer is used twice with the scan functions separated *in time* (described in Section 2.3.2).

2.3.7.1.1   Triple Quadrupole (QqQ)

The triple quadrupole consist of two quadrupole mass analyzers (Q1 and Q3), and a central section between Q1 and Q3 that is an rf-only component, designated with a lowercase q (q2) (Figure 2.34). The rf field in q2 acts to constrain the ions, enabling their transfer between the two analytical quadrupoles (*broad band pass* mode). The central cell (q2) is the location where a collision (reagent) gas, usually argon, is introduced to effect collision-induced dissociation (CID). The pressure of the collision gas is held constant, while fragmentation is optimized by varying the amplitude of the voltage used to propel the ions leaving Q1 and into q2. It is also possible to affect the degree of fragmentation by altering the gas pressure; however, for speed, convenience, and reproducibility, it is preferable to change the collision energy (voltage) while holding the gas pressure constant. The products of the CID process are analyzed in Q3 by scanning to collect full spectra or by recording the intensity of a

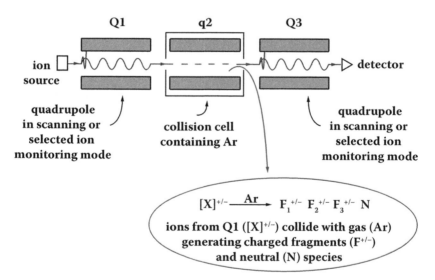

Q1 and Q3 are analytical quadrupoles that can be scanned to obtain spectra or held constant to select an ion of interest.
Ions that are unstable for a given setting of Q1 or Q3 are lost by collision and discharge on the quadrupole rods.
q2 is a transmission, rf only, cell into which a collision gas is introduced to effect CID (collision-induced dissociation).
The ion source may be internal (EI or CI), or external (API).

FIGURE 2.34   Principle of the triple quadrupole (QqQ) mass analyzer.

specific ion. Triple quadrupole instruments are remarkably versatile because the Q1 and Q3 analyzers can be used in conjunction with each other in either scanning or static (selected ion monitoring) mode (Section 3.3.3.1).

2.3.7.1.2 Time-of-Flight/Time-of-Flight (TOF/TOF)

A certain amount of structural information, such as sequencing of peptides, can be obtained from single-analyzer MALDI-TOF systems used in the reflectron mode because some ions leaving the ion source are *metastable* (Section 3.3.1.2) and fragment as they move down the flight tube; the process is called *post-source decay* (PSD). The technique is limited because it is not possible to preselect ions for fragmentation and because stepwise changes in the parameters of the reflectron are needed to obtain product ion spectra.

The combination of two analyzers in a TOF/TOF overcomes the problems with the detection of PSD ions. The separation attained with the first analyzer enables the selection of an ion of interest using an electronic gate called a *timed ion selector* (TIS). The selected ion is then fragmented in a collision cell, followed by analysis of the fragments in a second TOF analyzer (Figure 2.35). The calibration of TOF 1 (with known compounds) permits the prediction of the arrival time of the

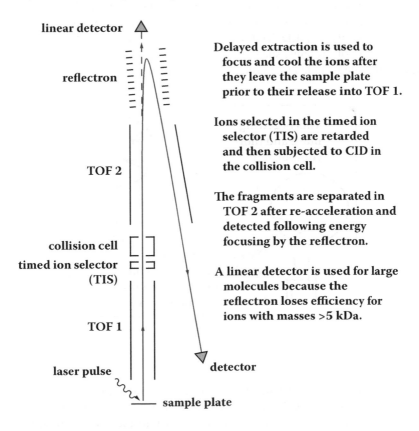

linear detector

reflectron

TOF 2

collision cell

timed ion selector (TIS)

TOF 1

laser pulse

detector

sample plate

Delayed extraction is used to focus and cool the ions after they leave the sample plate prior to their release into TOF 1.

Ions selected in the timed ion selector (TIS) are retarded and then subjected to CID in the collision cell.

The fragments are separated in TOF 2 after re-acceleration and detected following energy focusing by the reflectron.

A linear detector is used for large molecules because the reflectron loses efficiency for ions with masses >5 kDa.

FIGURE 2.35 Outline of a MALDI-TOF/TOF analyzer.

ion of interest at the TIS, where high-speed electronics are used to deflect other ions arriving before and after the analyte ion. Normal isotope distributions are observed for the fragment ions because the window within which the selected m/z lies is several mass units wide. Ions leaving the ion source with kV acceleration possess energies that are too high for CID in terms of both the extent of the fragmentation and the time spent in the collision cell. The selected ion must therefore be decelerated before entering the collision cell. The emerging fragment ions are collected at a second acceleration point and then released into TOF 2, which is equipped with a reflectron.

As noted earlier, the mass range of reflectrons is limited to an upper limit of ~5 kDa. This mass is usually sufficient for the analysis of tryptic peptides. At masses >5 kDa, a linear detector must be used and resolution drops rapidly to <100, with peaks becoming very broad. The bases of the peaks are several hundred mass units wide even for small proteins, such as myoglobin at m/z 16,950, (illustrated in Figure 2.20), and the accuracy of mass measurement is only ~0.1%.

MALDI-TOF and MALDI-TOF/TOF are straightforward techniques, and a large number of samples can be analyzed rapidly, particularly for masses in the 500 Da to 5 kDa mass range, e.g., to obtain mass maps (MS) and sequences (MS/MS) of tryptic peptides that have been spotted on plates robotically. However, the lack of resolution and accuracy are limiting when analyzing proteins in the linear mode. Accordingly, MALDI-TOF cannot be used to assess small changes in the molecular mass of proteins, such as those that are due to phosphorylation or when a small molecule is added to a protein.

2.3.7.2   Hybrid MS/MS Instruments

2.3.7.2.1   Quadrupole Time-of-Flight (QTOF)

QTOF systems (also called QqTOF) are versatile configurations. The quadrupole may be used in either a wide or narrow band pass mode to determine which ions are passed into the collision region (q) and then into the time-of-flight analyzer (Figure 2.36). Full-scan MS data are obtained in the basic mode when the quadrupole

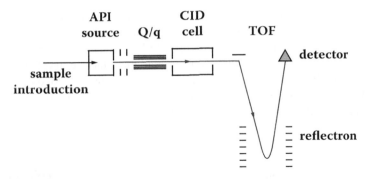

Ions are produced in an atmospheric pressure ionizaton source.
The quadrupole can be used in broad band pass (q) mode, to pass
  all ions to the time-of-flight (TOF) analyzer, or in narrow band pass
  (Q) mode to select an ion with a specific *m/z* for collision induced
  dissociation (CID).
The CID cell also has a q function by which the fragments formed
  are constrained and transfered to the TOF analyzer.
CID fragment ions are separated and collected in the TOF analyzer.

FIGURE 2.36   Principle of a QTOF MS/MS instrument.

analyzer is operated in the wide band pass format, there is no gas in the collision cell, and all ions are transferred into the TOF analyzer. When MS/MS data are required, the quadrupole is set in narrow band pass operating mode, which is particularly useful because of the ability to rapidly switch the quadrupole field, thereby enabling the individual selection of multiple ions for fragmentation, e.g., from within a single chromatographic peak. To fragment an ion of known m/z, Q is set to act as a filter (selection device), allowing passage of the desired m/z while all unwanted species are removed. Information on all other ions entering Q will be lost.

The inherent sensitivity, resolution, and rapid data acquisition of TOF analyzers are major assets of LC-MS/MS analysis. The TOF format provides accurate mass data on both molecular species and fragments. QTOF instruments feature multiple ionization modes, including convenient switching between ESI and MALDI.

Increased analytical sophistication can be attained through incorporation of an ion mobility separator (IMS) into the QTOF format, e.g., a Qq_1IMSq_2TOF system where the collision gas can be placed in either q cell. Such sophisticated instruments enable multiple types of experiments. For instance, a particular m/z that may consist of multiple species can be selected in Q, trapped in q_1, and passed into the IMS cell where components of the chosen m/z are separated according to their cross-sectional areas. Ions with different configurations emerge sequentially from the IMS cell and undergo CID in q_2, followed by collection of the resulting MS/MS spectra using the TOF analyzer.

2.3.7.2.2   Linear Ion Trap–Fourier Transform (FT) Analyzers
In FT-based hybrid instruments the first analyzer is a linear ion trap (LIT), while the second analyzer is either an orbitrap or an ICR cell. These analyzer combinations

enhance utility because the LIT is itself an MS/MS instrument that can be used both as an ion selection device and for collecting spectra. The trapping capability of the LIT makes MS^n analysis possible (usually up to MS^3), together with transfer of specific ions to the second, FT-based analyzer for accurate mass measurement (Figure 2.37). To utilize the highest resolutions on the FT analyzer (Section 2.3.4), data acquisition times of both the LIT and FT analyzers must be considered. The high data acquisition speed of the LIT allows the acquisition of the MS/MS spectrum of an m/z, while part

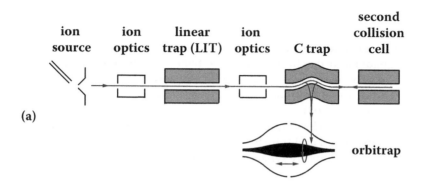

(a)

**The LIT-orbitrap combination can be used for both MS and
  MS^n experiments because the LIT has its own detector.
Additional fragmentation at a different collision energy can be
  undertaken in a second collision cell.
Products from either collision cell are injected, via the C trap,
  into the orbitrap where transients are collected for subsequent
  FT analysis.**

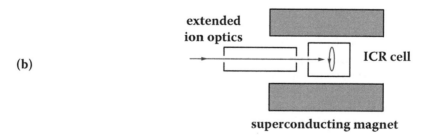

(b)

**In the LIT-ICRMS, ions from the LIT traverse a set of extended
  optics into the ICR cell located within a superconducting magnet.
A ramped rf voltage pulse is used to coalesce different m/z
  values and move them close to the reciever plates where
  transients are collected for subsequent FT analysis.**

FIGURE 2.37 Schematic of a hybrid linear ion trap with an orbitrap or ICR cell as the second analyzer.

of the same m/z is transferred into the FT cell for highly accurate mass measurement. The MS/MS data acquired in the LIT are only at nominal mass.

When a second collision cell is incorporated into an LIT-orbitrap combination, different energies are used in each collision cell to obtain additional structural information (Figure 2.37a). Extended optics are required in ICR systems to transfer ions emerging from the LIT into the second analyzer because the cell is located within a superconducting magnet (Figure 2.37b).

2.4   ION CURRENT DETECTORS

The role of the detector, properly called an *ion current* detector, is to determine the abundance of the ions of different m/z after they have been separated in, and emerge from, the mass analyzer. Detectors are also referred to as *multipliers* because they multiply (amplify) the very low current produced by the ions arriving at the detector surface. Ions traveling from the source through the analyzer to the detector constitute an ion current. The arrival of one ion per second corresponds to a current of 1.6×10^{-19} A. Ion currents in most mass spectrometers are in the 10^{-9} to 10^{-16} A range. As long as ions of a particular m/z continue to arrive at the detector, the small current can be amplified to increase the signal attributable to that m/z value.

Photoplate detectors are of historical significance because they were used (first by Thomson in 1911) to detect, for the first time, ions in a *parabola positive ray apparatus*. A photoplate consists of an emulsion of a suspension of silver bromide crystals in gelatin, spread as a thin film on a glass plate. A photosensitive emulsion is a composite collector-transducer-recorder. After energetic ions impinge on the surface, a chemical amplification occurs and a latent image is formed on the emulsion. Masses and their abundances are determined from the positions and the intensities (blackness) of the lines. Photoplates were used for *multicomponent* detection in double-focusing magnetic analyzers where all resolved ion beams are focused simultaneously along a focal plane. Advantages include the ability to record, simultaneously, and at high resolution, hundreds of ion beams of individual m/z values. Disadvantages include low sensitivity (10^4 ions required for detectable lines), nonlinear response, and a need for expensive densitometers. Photoplates have been replaced by arrays of individual detectors that act as electronic photoplates.

2.4.1   ELECTRON MULTIPLIERS

The first electronic ion detector was the *Faraday cup*. It consists of a metal container in which ions arriving from the mass analyzer are discharged and generate a current. While it is simple and capable of very precise operation, the Faraday cup is used only in specific applications where sufficient sample is available because it cannot amplify the signal derived from the very small charge carried by each arriving ion; i.e., the cup is a detector but not a multiplier.

A combination of amplification steps is employed to improve detection sensitivity. First, the signal is amplified within the detector, and then there is subsequent electronic amplification, outside the vacuum chamber, followed by recording by the data system. Detectors that amplify the current associated with the arriving ions (*electron multipliers*) come in a variety of forms, including the *discrete dynode* electron multiplier, the *continuous dynode* electron multiplier, and the *multichannel plate* detector. The *Daly* detector relies on the production of photons by ions incident on a surface coated with a scintillant, followed by amplification using a photomultiplier (Section 2.4.2).

Although the structures of different electron multipliers are often significantly different, the principle of their operation is essentially the same (Figure 2.38). The first step is to convert the ion current into an electric current, i.e., one that is composed of electrons instead of ions. This is accomplished by accelerating the ions arriving from the mass analyzer onto a *conversion dynode* where collisions release electrons that are in turn accelerated into the multiplier. A voltage gradient across the components of the multiplier produces a cascade of electrons. Multipliers generate current gains that are on the order of 10^6. There is no conversion dynode in multichannel plate (MCP) detectors where the electron cascade is initiated directly by the arriving ions.

The oldest electron multiplier is the discrete dynode, where a series of copper/beryllium plates act as mirrors reflecting the electrons, initially ejected from the conversion dynode, from one plate to the next, sequentially releasing more electrons from each plate thereby amplifying the signal at each step (Figure 2.39a). The electron cascade from mirror to mirror is potentiated by a voltage gradient along the length of the multiplier. A later version, the continuous dynode system, facilitates the electron cascade within a curved glass funnel where a voltage gradient is established on a resistive surface. Electrons entering the funnel cascade and amplify the incoming signal (Figure 2.39b).

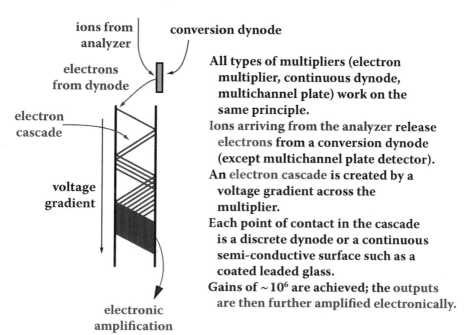

FIGURE 2.38   Principle of electron multiplier detectors.

discrete dynode

(a)

The discrete dynode multiplier
is composed of a series of
copper:berylium plates between
which the electrons cascade.

continuous dynode

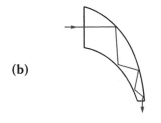

(b)

The continuous dynode is a
leaded glass funnel coated on
the inside with a semi-conductive
material.

**multichannel
plate**

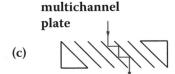

(c)

The multichannel plate detector
(MCP) is a porous glass plate in
which each pore acts as a mini
electron multiplier. The MCP
is a set of parallel multipliers.

The three varieties of detector shown have different structures
but are based on the same principle.
The discrete and continuous dynode versions are used
on scanning instruments. Ions are directed to a conversion
dynode to generate electrons that are accelerated to the
multiplier.
The MCP system is used for time-of-flight instruments. The
ions arrive at the plate and initiate the electron cascade.

FIGURE 2.39 Types of electron multipliers.

The discrete and continuous dynode detectors work well with both quadrupole
and magnetic sector instruments. However, they do not provide the rapid detection
and amplification required by TOF analyzers. MCP detectors (Figure 2.39c), where
the electron cascade occurs within the individual pores of a porous glass plate, are
fast enough for TOF analyzers. In essence, each pore (~5 microns in diameter) is a
micromultiplier. Detection by MCP is limited because the plates can receive only one
ion of a given empirical formula within a specific time frame because the *time-to-
digital converters* (TDCs), which acquire the signal from the MCP, have a recovery
period of nanoseconds (called the dead time) before another signal can be processed.
This delay is not an issue for ions of different *m/z* as they are sufficiently separated in
time. However, when two ions of the same empirical formula reach the MCP, only the
one arriving first will be recorded. The dynamic range of MCP detectors is limited
(two to three orders of magnitude) because the dead-time problem leads to their rapid

saturation; thus, MCP detectors are not suitable for quantification. The saturation also causes distortion of accurate mass measurement because as ions of the same empirical formula arrive at the detector, only those that have slightly higher energies (and therefore are traveling faster) will be detected. These ions will appear to have slightly lower masses because measured mass is a function of arrival time at the detector.

The latest generation of analog-to-digital converter (ADC) multipliers for TOF analyzers improves the speed of data acquisition and processing significantly. These detectors consist of a flat surface that acts as a conversion dynode from which the emitted electrons are guided to the multiplier using a magnetic field. The new ADC multipliers, which provide greater dynamic range than TDC systems, have extended the utility of TOF analyzers in quantification and the concentrations over which accurate mass measurements can be made.

The lifetime of electron multipliers is 1–2 years because contamination of the surfaces reduces the number of electrons released. Multipliers often work adequately for several years because in the majority of applications there is no need for maximum sensitivity.

2.4.2 Daly Detector

The Daly detector uses a photomultiplier rather than an electron multiplier. Ions leaving the analyzer are directed onto a conversion dynode, and the ejected electrons are accelerated onto a plate coated with a fast-acting scintillant. Each electron releases a photon from the scintillant. The photons then enter a photomultiplier tube and impact on a photocathode, producing electrons (photoelectric effect) and initiating an electron cascade (Figure 2.40). The output from the photomultiplier is further amplified electronically, similarly to the output of dynode type electron multipliers. The level of amplification is similar to that of electron multipliers. Photomultiplier tubes last longer than electron multipliers, but the scintillant-coated plates require replacement every few years.

2.4.3 Image Currents

The FT instruments do not need electron multipliers. Instead, ion detection is based on the phenomenon that the passage of ions close to a metal surface induces an electric current in that surface, similarly to the action of the armature in electrical generators. The result is an alternating current with a sinusoidal path the frequency of which is proportional to the m/z of the ion that induced the current. FT systems are nonscanning instruments, thus all ions in the analytical cell are coalesced, using a ramped rf voltage, into packets corresponding to their m/z and then detected simultaneously. Each ion packet produces its own sinusoidal current with a frequency that is proportional to its m/z value. The final image current, a highly complex product of overlapping, additive, sinusoidal curves, must be deconvoluted mathematically, using Fourier transform (FT) techniques, to determine the m/z values of the constituents.

When two sinusoidal curves interact, the resulting complex wave contains a harmonic that reflects the frequencies of the two curves. The periodicity of the harmonic is much longer than the frequencies of the signals from which it is derived.

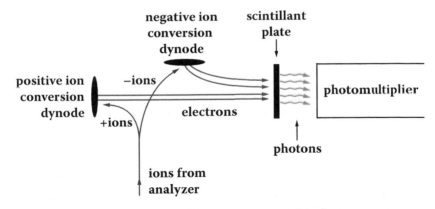

Ions (positive or negative) impact the polarity-specific
conversion dynode and electrons are ejected.
The electrons are accelerated onto a plate coated with a fast
acting scintillant with subsequent emission of photons.
The photons are detected and the signal amplified by a
photomultiplier.

FIGURE 2.40   Principle of the Daly detector.

Furthermore, the closer together the frequencies, the longer is the time function for
the interaction harmonic. A highly complex set of harmonics is produced when multiple sinusoidal curves are superimposed. The m/z values are calculated from the
periodicities of the interaction harmonics derived from FT analysis of the acquired
transient. The need to obtain the harmonic frequencies is the reason for the extended
time periods over which signals from FT instruments must be acquired; the closer
together the constituent m/z values, the longer is the periodicity of the harmonic
interaction that must be detected.

By way of an analogy, think of tuning a musical instrument where one listens for
the pulsed harmonic that occurs between two notes that are not quite the same. The
periodicity of the harmonic increases as the notes are brought into tune with each
other. The harmonic is measured in fractions of a second (Hz), much longer than are
the frequencies of the constituent sounds, e.g., the frequency of middle C is 262 Hz.
This is an example of the type of harmonic that can be detected using Fourier transform analysis.

A particular advantage of FT instruments is that the variations in the image current generated can be measured continuously and precisely, providing an uninter-
rupted, continuous profile of events in the analyzer. This is in contrast to conventional
mass spectrometers where the detection of ions is based on discrete events, namely,
the impact of ions on a surface and the subsequent amplification of the signal. FT
analysis is a mathematical process that can work on a continuous basis because the
transients are processed in infinitely small slices, thereby improving the quality of
the information obtained. In FT-based systems the generation and processing of data

are complementary, as both are continuous, rather than the consequence of individual events, as occurs with ions arriving at electron multipliers.

2.5 VACUUM SYSTEMS

The vacuum systems used in mass spectrometers are intended to prevent the loss of ions by collision with neutral molecules, such as air, in the various chambers of the instrument. Another important function is to remove unreacted molecules from the ion source to prevent/reduce *memory effects* by minimizing cross-contamination between samples introduced in rapid succession, such as from successively eluting components in chromatographic or electrophoretic effluents. However, there are occasions when collisions between ions and neutral molecules are desirable, such as in collision-induced dissociation and for the collisional cooling of ions. Thus, it is important to consider the degree of vacuum needed in various parts of mass spectrometers.

Pressure is a force acting on a unit area. In mass spectrometry, a unit of pressure commonly used is the Torr, where 1 Torr is equivalent to approximately 1 mm of mercury, and 760 mm of mercury is 1 atmosphere (atm). A more accurate definition, which accounts for the variation in the density of mercury, is that a Torr is 1/760 of an atmosphere, which is itself defined as 101,325 pascals (Pa). There are a number of other units in use, and this can (and often does) cause confusion; e.g., while Torr is typically used in the United States, millibar (1/1,000 atm) is used in Europe and pascal in Japan.

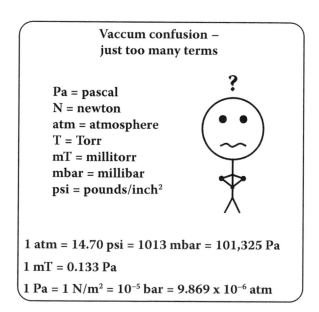

**Vaccum confusion –
just too many terms**

Pa = pascal
N = newton
atm = atmosphere
T = Torr
mT = millitorr
mbar = millibar
psi = pounds/inch²

1 atm = 14.70 psi = 1013 mbar = 101,325 Pa

1 mT = 0.133 Pa

1 Pa = 1 N/m² = 10⁻⁵ bar = 9.869 x 10⁻⁶ atm

Changes in pressure are often given in orders of magnitude, e.g., 10^3 and 10^{-4}. Because pressure can be expressed as positive or negative, there is, in fact, no need for the term *vacuum*; however, in common usage, negative pressure is called vacuum. The use of the two terms is confusing; e.g., a pressure of 10^{-7} Torr is both a low pressure and a high vacuum. In mass spectrometers vacuum is also described in still different terms, reflecting the extent to which a chamber has been evacuated, including rough vacuum (ranging from atmospheric pressure to ~10^{-2} Torr), high vacuum (from 10^{-4} to 10^{-7} Torr), and ultra-high vacuum (10^{-7} to 10^{-10} Torr).

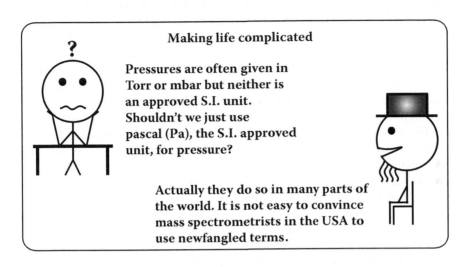

Making life complicated

Pressures are often given in Torr or mbar but neither is an approved S.I. unit. Shouldn't we just use pascal (Pa), the S.I. approved unit, for pressure?

Actually they do so in many parts of the world. It is not easy to convince mass spectrometrists in the USA to use newfangled terms.

The lifetime of an ion in a mass spectrometer depends not only on whether the ion is stable when energy is added during ionization, but also on whether the ion is lost though collision with a molecule (e.g., an air component) or a surface. Such events may reduce resolution and sensitivity significantly. The average distance that an ion can travel before encountering a neutral molecule (at a specified pressure) is called the *mean free path* (λ). Vacuum systems are designed to maximize λ to reduce ion losses. Table 2.2 provides numerical values of mean free paths and the operating pressures for different mass analyzers.

TABLE 2.2
Required Vacuum and Mean Free Path for Various Mass Analyzers

Analyzer	Mean Free Path (λ)	Pressure		
		Torr	mbar	Pa
	5 cm	10^{-3}	1.3×10^{-3}	1.3×10^{-1}
Q, QIT	50 cm	10^{-4}	1.3×10^{-4}	1.3×10^{-2}
Q, QIT	500 cm	10^{-5}	1.3×10^{-5}	1.3×10^{-3}
TOF, magnetic	5,000 cm = 5 m	10^{-6}	1.3×10^{-6}	1.3×10^{-4}
TOF, magnetic	50 m	10^{-7}	1.3×10^{-7}	1.3×10^{-5}
	500 m	10^{-8}	1.3×10^{-8}	1.3×10^{-6}
	5,000 m = 5 km	10^{-9}	1.3×10^{-9}	1.3×10^{-7}
ICR, orbitrap	50 km	10^{-10}	1.3×10^{-10}	1.3×10^{-8}

The length of the mean free path and, consequently, the required vacuum is related to the overall size of most instruments, e.g., 10^{-5} Torr for a quadrupole. The vacuum requirement of 10^{-10} Torr for the ICR and orbitrap may seem surprising at first glance. Consider: If the orbit that an ion follows in an ICR cell has a diameter of 3 cm, the circumference of the orbit is $2\pi r$ or ~10 cm; given that the orbital frequency of an ion of m/z 400 is ~350,000 Hz (in a 9.4 Tesla magnet), the distance that an ion travels per second is 350,000 Hz × 10 cm = 3,500,000 cm = 3.5 km (over 2 miles). Obviously, a vacuum of 10^{-10} Torr is required to avoid unwanted collisions.

Obtaining vacuum in a mass spectrometer is composed of two stages. The first of these provides an initial (rough) vacuum that takes the system from atmospheric pressure down to ~10^{-2} Torr (1.3 × 10^{-2} mbar, 1.3 Pa). In the second stage, high-vacuum pumps evacuate the various chambers of the instrument to the required 10^{-5} to 10^{-10} Torr (1.3 × 10^{-5} to 1.3 × 10^{-10} mbar, 1.3 × 10^{-3} to 1.3 × 10^{-8} Pa). The high-vacuum pumps begin to operate only once a pressure (vacuum) of ~10^{-2} Torr is reached.

An important requirement of the vacuum systems of modern mass spectrometers is to provide *differential pumping* of various chambers that are often separated by apertures with low conductance. The necessary vacuum is obtained using independent pumps on the individual chambers (e.g., source and analyzer). Differential pumping is needed, for instance, because the high pressure of the reagent gas in CI sources must not be allowed to affect the analyzer region; thus, each area is pumped separately. A small orifice between the source and analyzer enables efficient passage of the ions but limits the pressure equalization between the two chambers. Other examples of locations where considerable gas pressure must be present are the cells used for CID and IMS, where fast and efficient removal of the added gas is essential. Differential pumping may be needed even within the same component; e.g., in ESI sources there are several regions where the pressure must be stepped down between the point where the liquid spray is generated and the entrance of the ion beam into the analyzer.

2.5.1 BACKING (ROUGHING) PUMPS

The backing pumps serve two purposes: (1) to quickly remove atmospheric gases and establish the minimum vacuum required for the operation of high-vacuum pumps, and (2) to remove the exhaust gases coming from the high-vacuum pumps. Rotary-vane and scroll pumps are the common types of backing pumps.

2.5.1.1 Rotary Pumps

The oil-based rotary pumps, in use for decades, remain the most common way to generate rough vacuums (Figure 2.41a). Their mode of operation is based on an off-set rotor revolving in a cylindrical chamber, while sprung flaps (vanes) on the rotor maintain contact with the cylinder walls. The offset rotor-vane combination sweeps a packet of captured gas around the pump housing, compressing the gas as it is pushed toward an exhaust outlet. The increased pressure causes the expulsion of the gas through the exhaust port (Figure 2.41b). One disadvantage, particularly when using API, is that the oil in the pump connected to the ion source must be changed frequently because of contamination with the solvents used during sample introduction, e.g., the water, methanol, and acetonitrile used in reversed-phase chromatography.

(a) **Layout of a rotary vacuum pump**

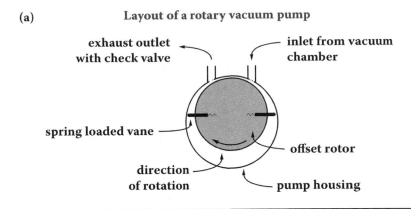

(b) **Motion of rotor and vanes to generate a vacuum**

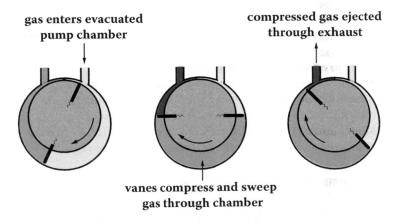

As the rotor turns, the vanes remain in contact with the pump
housing forming a moving seal. Pockets of gas, captured at the
pump inlet, are swept through the chamber and is compressed as
it nears the exhaust outlet thus creating a positive preasure that
ejects the gas through the outlet, past a check valve.

FIGURE 2.41 Principle of rotary vacuum pumps.

Another problem is that *back streaming* of the lighter components of the pump oil
can contaminate both the ion source and analyzer; e.g., contamination in EI sources
results in the appearance of characteristic hydrocarbon fragments, while contamination of analyzers leads to loss of resolution and sensitivity.

2.5.1.2 Scroll Pumps

The operation of scroll pumps, a new generation of backing pumps, is based on two
Archimedian spirals, one fixed while the other moves with a slightly offset rotation.
An Archimedian spiral is one where the distance (spacing) between each rotation of
the spiral is constant. Similarly to rotary pumps, scroll pumps operate by capturing

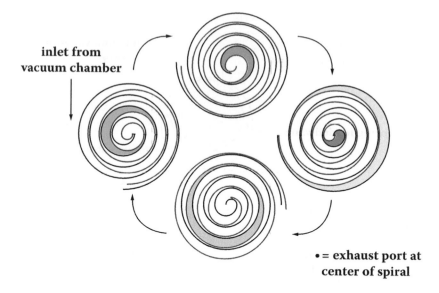

inlet from
vacuum chamber

• = exhaust port at
center of spiral

There are two offset Archimedian spirals.
The moving spiral (red) pushes a pocket of evacuated air
   through the spiral pathway thereby compressing the air
   until it escapes through the central exhaust vent.
Movement of the air is illustrated in each quadrant with the
   compression shown by the intensity of the blue regions.

FIGURE 2.42   Principle of scroll pumps.

a pocket of gas and compressing it into a progressively smaller volume and eventually expelling the gas through an exhaust port. The gas is compressed as it moves
through the steadily decreasing volumes that exist toward the center of the spiral.
The exhaust port is at the center of the spiral (Figure 2.42). Benefits include that
no oil is used, thus reducing contamination, and the moving surfaces are made of
Teflon that can be readily replaced during maintenance. Although more expensive
than rotary pumps, scroll pumps are becoming popular; added benefits are reduced
generation of heat and less noise.

2.5.2   TURBOMOLECULAR PUMPS

The high vacuum required for instrument operation is usually provided by turbomolecular pumps. A turbo pump is basically a sophisticated fan spinning at very high
speeds, 50,000 to 90,000 rpm. To imagine the inside of a turbo pump, think about
looking into the front of a jet engine. The view is of a set of fan blades that pull in
outside air, compress it, and then expel it into a combustion chamber. If, instead of
pulling in the outside air, the fan is attached to a sealed chamber, then vacuum will
be created as the evacuated air is compressed and forced through the pump, where it
is removed as exhaust by a backing pump. Turbo pumps are comprised of alternating

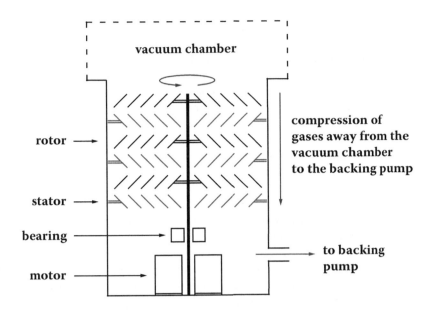

Turbomolecular pumps are based on the blades of the
high speed rotors (50,000 – 90,000 rpm) providing
directional acceleration to the gas molecules, moving
them away from the chamber that is being evacuated.
The moving gas molecules are deflected off the stator
blades into the next rotor with eventual removal
by the backing pump.

FIGURE 2.43 Principle of turbomolecular pumps.

rotors and stators (Figure 2.43). The angled rotor blades hit the gas molecules, driving them into the pump and away from the chamber that is being evacuated. Passage through the stator sets up the flow to the next set of rotor blades. Several (~8) such rotor:stator combinations are stacked together, with each pairing providing ~10-fold compression as gases are forced away from the chamber.

Turbomolecular pumps have replaced the previously used diffusion pumps because they do not use oil, thereby preventing contamination, and have short start-up and shutdown times. Modern turbomolecular pumps often run for several years before mechanical failure occurs from wear on the bearings.

2.6   DATA SYSTEMS

Major advancements in mass spectrometric instrumentation (e.g., ESI and MALDI) and electronics (e.g., TOF data handling) have been accompanied by rapid progress in computerization. Older mass spectrometrists may still remember when hand-adjusted potentiometers were used to manipulate the ion beam, and spectra were collected on photographic paper where the peaks had to be counted by hand (using background ions to determine the mass scale) as well as when the appearance of small, programmable calculators was considered a major advance. Today, computers of widely varying sophistication are dedicated, integral, and essential components of mass spectrometer systems. The term *data system* refers to the combination of computer hardware together with often astonishingly sophisticated software, used in all aspects of mass spectrometry. This is true, even for the simplest instruments employed for straightforward tasks.

Dedicated data systems perform four functions: (1) control all operational processes of both the mass spectrometer and integrated peripheral instruments, such as GC or LC systems; (2) acquisition and processing of all data; (3) local interpretation of acquired data; and (4) post-processing of data, including interaction with databases (almost always via the Internet) (Figure 2.44). Connection to the Internet also enables the remote control of multiple systems as well as the off-site diagnosis of failures by instrument manufacturers.

The operation control functions range from relatively simple tasks, such as instrument "tuning" (e.g., automatic control of the ion beam and mass calibration), to highly sophisticated automatic switching from one mode of operation to another, such as data-dependent acquisition in the MS/MS mode triggered by ions (obtained

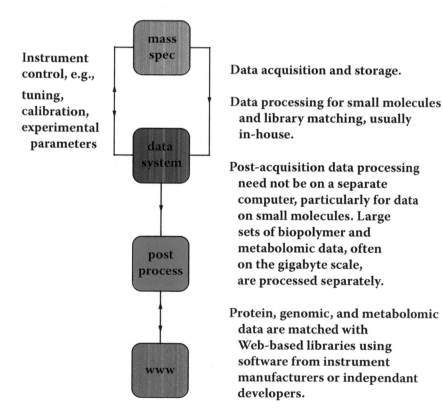

FIGURE 2.44 Computing and data handling.

in the MS mode) with particular *m/z* values and abundances above a preselected threshold (Section 3.3.3.4).

The data acquisition function includes the digitization and processing of the electrical signals from the ion detectors to provide some form of conventional mass spectra and storage of the resulting data. The processes (conversions) cover a range of signal manipulations (often automatic) such as *thresholding*, which controls the points at which a signal is considered an *m/z* peak, and *noise reduction* (smoothing), which reduces the effects of electronic noise on the acquired data. Other tasks in this category include *centroiding*, by which a Gaussian profile data peak is converted into a single line corresponding to the geometric center and intensity of the peak, and the *deconvolution* of multiply charged spectral envelopes for analytes of high molecular mass produced by ESI. More sophisticated data processing techniques include the interpretation of the image currents (by Fourier transform analysis) obtained in orbi-traps and ICR instruments.

Software packages for instrument control usually also include programs to handle a wide variety post-acquisition data tasks, such as searching local (as well as Web-based) EI databases for analyte identification (small molecules), quantification of drugs and metabolites, the necessary forms and data arrangement to meet regulatory reporting requirements (environmental contaminants), peptide sequencing, and comparing sequence spectra from different samples to search for the presence of unique biomarkers or differences in their concentrations.

There has been a remarkable proliferation of data handling software for proteomic studies offered by instrument manufacturers as well as by independent software developers. The Internet has allowed the relatively facile uploading and storage (with impressively frequent updating) of extensive data libraries of protein sequences and genomic information on a species-by-species basis, with access available without charge. For example, a given set of mass spectra can be compared against the proteomic/genomic databases to identify individual proteins and the species from which they have been isolated. Software tools also provide *probability scores* for the fit of the spectra to the database information, whether there are other proteins that show sequence similarities, possible post-translational modifications, etc. There are also searchable databases on the Internet for most currently fashionable *omics* type disciplines (Section 3.4), such as metabolomics.

Modern commercial mass spectrometers are so reliable that they can often be operated unattended on a 24/7 basis. The acquired data sets can reach gigabytes per sample. Such large data sets are moved off the instrument control computer for processing, and until a few years ago, separate, multicomponent computer formats (often off-site) were required for processing and interpretation. The present generation of computers with fast, multicore processors, extensive RAM, and terabyte hard drives often permits the undertaking of most operations on single computer systems.

It is noted in passing, that the almost inconceivable variety and sophistication of the available software, particularly in the areas of interpretation (and often presentation) of mass spectral data remind one of the capabilities of Word, Excel, the apps for iPad, and similar consumer products: while an individual often uses only a small percentage of the available capabilities, it is reassuring to know that they are always available when the need arises.

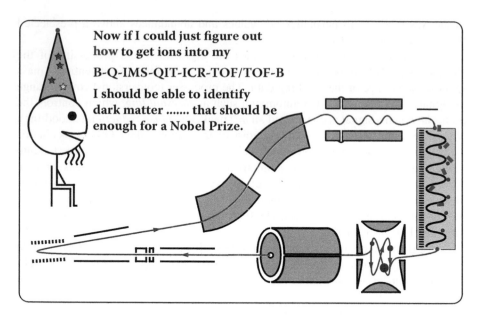

3 Methodologies and Strategies

This chapter describes and discusses the measures of instrument performance, the types of information that can be acquired, and the methodologies and strategies available. The discussions are with respect to both the multiple types of instruments manufactured and the various classes of compounds that can be analyzed. The range of compounds makes defining the types of data that can be obtained somewhat problematic. Therefore, small molecules are dealt with in Sections 3.2 and 3.3, while biopolymers are covered in Section 3.5, although the differentiation is somewhat artificial. For instance, the determination of accurate mass data described in Section 3.1.3 is relevant to both small molecules and the analysis of the peptides derived from proteins.

3.1 MEASURES OF INSTRUMENT PERFORMANCE

3.1.1 INSTRUMENT TUNING AND MASS SCALE CALIBRATION

All mass spectrometers require some degree of day-to-day tuning of operational parameters to provide optimal peak shapes, resolution, and sensitivity, as well as calibration of the mass scale. These aspects are important for the characterization of molecules and their fragments, the resolution of one compound from another, and the quantification of an analyte.

The instrument software usually provides automated routines for tuning, but manual intervention may also be required to obtain the best performance. One objective of tuning is to make the minimum necessary adjustments to the transmission of the ion beam to reduce ion losses and to provide high sensitivities. Tuning should begin at the ion source and proceed in sequence, progressively optimizing the voltages that guide the ions from the source to the analyzer and onto the detector. In addition, voltages and gas flows for the API methods also need tuning as these parameters affect the entry of ions into the vacuum chamber. There is a variety of standard compounds used for tuning and mass scale calibration (Table 3.1).

Mass calibration establishes the mass scale by determining the relationship between a series of m/z values, covering the mass range of interest, and the operational parameters that determine how a mass spectrum is collected, e.g., scanning of voltages on a set of quadrupole rods, or measurement of the flight times of ions in a TOF analyzer. Figure 3.1 illustrates a series of calibration ions as well as ions for an unknown and a lock mass.

The stringency and frequency with which calibration procedures should be undertaken is determined by whether an instrument is being used for nominal or accurate mass measurement. Modern instruments are very stable, and the mass scale

TABLE 3.1

Typical Compounds for Calibration/Tuning

Calibration/Tuning Compound	Ionization Method and Mass Range
Perfluorotributylamine (FC-43)	EI 1–650
Perfluorokerosene (PFK)	EI 1–1,000
Triazines and Ultramark family	EI 1–1,000 (2,000)
Polyethylene glycol	APCI/ESI 100–2,000
Sodium formate (forms clusters)	ESI 100–1,500 (positive, negative ion)
Sodium, rubidium, cesium iodide (form clusters)	ESI 100–5,000+ (positive, negative ion)
Phosphazines	ESI 100–3,000
Peptides (e.g., bradykinin)	MALDI 500–5,000
Peptides and proteins	MALDI 1,000–150,000+

Mass scales are calibrated by introducing a known series of compounds (or one compound with known fragments) and fitting an equation that establishes the relationship between the acquired m/z values while also taking into account parameters such as scan rate.

To obtain the nominal mass (+/– 0.5 Da) of an unknown it is sufficient to match the observed m/z to a mass scale determined by a calibrant.

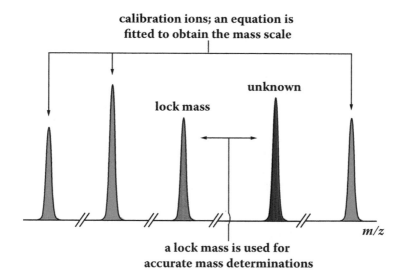

FIGURE 3.1 Mass scale calibration and the use of a lock mass.

For accurate mass measurements the unknown m/z are determined with respect to a lock mass to account for both *intra*- and *inter*-run variability.

The lock mass does not have to be one of the calibrant ions, but it has to be a compound with a known empirical formula.

may remain constant for months for the generation of nominal masses, up to ~3 kDa (150 kDa or more for MALDI-TOF). Matching experimental results with an existing mass scale calibration is a reliable method for obtaining the nominal masses of unknown analytes or their fragments.

Beyond nominal mass data, significantly more information is provided by using *accurate mass measurement*, where the masses of ions (*m/z* values) are measured to the third or fourth decimal place. To obtain this level of accuracy, the mass calibration has to be improved, for instance, by using numerous data points to more precisely define the mass scale. However, *intra-* and *inter-run* variations are inevitable even with frequent, careful calibration. Such errors may be counteracted by including a known compound in the analysis, a so-called *lock mass*, which is introduced separately or together with the sample. The lock mass is then used to make minor adjustments to the calibration to account for variabilities in instrument performance. In GC-MS, the compound providing the lock mass ion is bled into the ion source through a batch inlet, while for electrospray ionization (ESI) a second probe can be used to introduce a lock mass independently. Figure 3.1 shows a series of calibration ions used to establish a mass scale, as well as a peak for a known lock mass to adjust the calibration and to enable determination of accurate mass data for unknown analytes.

When analytes of high mass, such as larger peptides or intact proteins, are analyzed using ESI, conventional calibration files are still valid to ~3 kDa, because the *m/z* values of the multiply charged ions formed are usually less than 3,000. With MALDI-TOF, however, ions typically carry a single charge, and therefore the mass scale must be calibrated to *m/z* 150,000 or higher by analyzing a mixture of peptides and proteins of known molecular mass distributed across the desired mass scale. MALDI instruments must be operated in linear mode above 5 kDa and resolution rapidly drops to 100 or less. The lack of resolution makes it difficult to obtain calibration data that are sufficiently accurate to establish the masses of unknown analytes, limiting the measurement of molecular masses to ±0.1%. However, when the reflectron mode is used (<5 kDa) mass calibration can be sufficient to enable accurate mass measurement.

3.1.2 Resolution

As defined in Section 1.4, the resolution of a mass analyzer is a measure of its ability to separate one mass from another. In defining resolution, the abbreviation FWHM stands for full width at half maximum (height), i.e., the *m/z* of the ion of interest divided by the width of the ion at 50% of its height (intensity). Thus, for a singly charged ion of mass 600 that has a width of 0.075 at half height, the resolution is 8,000 (600/0.075 = 8,000).

The order of increasing resolution (including typical values) for the different types of analyzer is quadrupole (unit) < time-of-flight (to 60,000) < orbitrap (to 250,000) < FT-ICRMS (to 3,000,000), where unit is the ability to resolve mass X from mass X+1 throughout the mass range (Figure 3.2). Associated with resolution, and shown in the figure, is the term *peak capacity*, the number of peaks that can be fitted in between two adjacent masses. As resolution increases so does peak capacity.

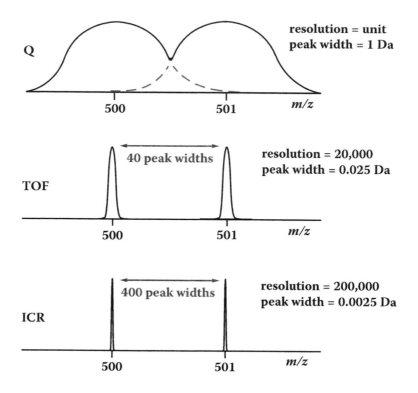

Resolution defines the ability to separate one *m/z* from another.
Peak capacity is the number of peak widths that can be
   accomodated between two unit masses. In this illustration
   there would be 40 peak widths on a TOF analyzer and 400
   on an FT-ICR system.

FIGURE 3.2   Examples of resolutions achievable with different analyzers.

When considering the resolution of an instrument, it is important to be aware of
the mass at which the resolution is measured and also the type of analyzer used. For
instance, for a TOF analyzer it is easier to attain a resolution of 8,000 at *m/z* 600 than it
is at *m/z* 200 because the peak width of an ion at *m/z* 200 must be one-third that of an
ion at *m/z* 600 to obtain the same resolution. It is harder to acquire the necessary data
points needed to achieve an accurate profile of a peak for an ion at *m/z* 200 ion at 8,000
resolution, where the peak is 0.025 *m/z* wide, than for an ion at *m/z* 600 ion, where the
peak is 0.075 *m/z* wide. The reverse is true for FT instruments where it is harder to
maintain resolution when moving up the mass scale. For example, an FT-ICRMS, with
a 9.8 Tesla magnet and using a 0.5 s transient acquisition, has a resolution of nearly
1,000,000 at *m/z* 40, but resolutions of 47,000 at *m/z* 800 and only 4,700 at *m/z* 8,000.

The resolution required for the separation of isobaric ions increases as one moves
up the mass scale because both the number of possible empirical formulae and the
number of possible elements increase. The situation is straightforward for mass 28,

as the only possible empirical formulae are CO, N_2, and C_2H_4, with exact masses of 27.9949, 28.0062, and 28.0313 Da, respectively. The difference between the masses of CO and N_2 is 0.0113 Da, while it is 0.0364 Da between CO and C_2H_4. A resolution of ~1,100 is needed to distinguish between N_2 and C_2H_4, while the resolution required for the separation of CO from N_2 is ~2,500, because these masses are closer together. Much higher resolutions are required to separate different empirical formulae at higher masses. For example, there are over 19,000 possible formulae for mass 500.1 ± 0.5 Da, even when the composition is restricted to $C_{10-100}H_{10-100}N_{0-10}O_{0-10}P_{0-5}S_{0-5}$. At the same time, there are only 28 possible formulae for 500.1 ± 0.0005 Da (0.5 mDa). Resolution of over a million would be required to resolve these 28 species. In other words, high resolution, while a very valuable asset in restricting the number of possible empirical formulae, cannot solve all problems.

3.1.3 ACCURATE MASS MEASUREMENT

The *accuracy* with which the mass of an ion can be determined is a combination of the ability to resolve one ion from another and how accurately the mass scale has been calibrated. There are multiple elemental compositions possible for any given nominal mass, e.g., (as noted above) there are three possible formulae, CO, N_2, and C_2H_4, at m/z 28, and the number increases rapidly as one moves up the mass scale. To obtain accurate measurements for each component of a set of isobaric ions, it is essential to resolve them from each other. As resolution increases and peaks become narrower, it becomes easier to be certain that an observed peak is a unique species and that the outline of the peak is Gaussian. Consequently, the center of gravity of the peak can be determined and the accurate mass of the ion calculated (Figure 3.3).

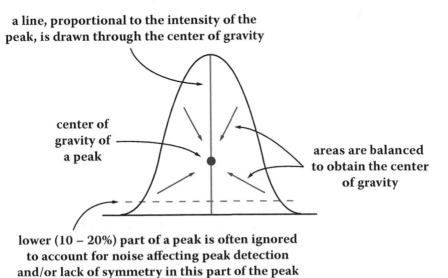

a line, proportional to the intensity of the peak, is drawn through the center of gravity

center of gravity of a peak

areas are balanced to obtain the center of gravity

lower (10 – 20%) part of a peak is often ignored to account for noise affecting peak detection and/or lack of symmetry in this part of the peak

FIGURE 3.3 Conversion of a peak from profile (analog) to centroided (stick figure) format.

Atomic masses are often thought of as unit numbers, e.g., carbon = 12, hydrogen = 1, nitrogen = 14, and oxygen = 16. However, only the carbon-12 isotope (^{12}C) is officially defined as a whole integer, 12 Da. The masses of all other elements are expressed with respect to this standard. Thus, hydrogen becomes 1.007825 Da, nitrogen = 14.0031 Da, and oxygen = 15.9949 Da. In fact, only elements up to and including nitrogen have fractional masses above their unit values; all others have masses below unit values (except thorium and uranium), with the 116 isotope of tin having the lowest exact mass at 115.9017 Da. Figure 3.4 illustrates how the variation in the fractional masses (also called mass defects) is a function of the atomic masses of the elements. The additional mass is termed *mass sufficiency* for elements with masses above a unit number, while for elements with masses below a unit number the variation is called *mass deficiency*. Remember that the exact mass value of an element always refers to a specific isotope of that element, and that the most important one is often the most abundant of its isotopes, e.g., ^{12}C, ^{28}Si.

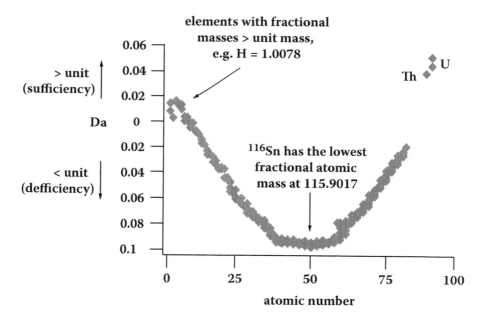

Graph of the differences of the elements from their unit mass values. Fractional masses are >0 up to nitrogen. For all other elements (except Th and U) the exact masses are below their unit mass values.

Consideration of these values provides ideas on the presence of certain elements when the mass of an unknown analyte is determined, e.g., incorporation of an iodine significantly lowers the fractional mass of a compound.

FIGURE 3.4   Variations in the fractional masses of elements.

A practical application of the value of resolution and accurate mass determination is shown in Figure 3.5. When the sulfa antibiotic sulfadimethoxine (MW 310) is ionized with ESI, the predominant ion is the protonated molecule at m/z 311. Fragmentation of this ion by collision-induced dissociation (CID) produces an ion at m/z 156. This ion appears as a single species when using a quadrupole analyzer; however, two isobaric species are observed at high resolution (Figure 3.5). A resolution of ~5,000 is required to fully separate the peaks. Accurate mass measurements show that the two ions at m/z 156 actually occur at m/z 156.0120 and m/z 156.0774, corresponding to the fragments representing the two sections of the molecule generated by breaking the –S–N– bond.

A few comments on terminology (see also Section 1.2): *Exact mass* is the sum of the known masses of the constituent atoms, whereas *accurate mass* refers to the experimentally determined value obtained using a mass spectrometer. The latter reflects the capabilities of the particular instrument used. These terms should not be confused with *molecular mass* (or *atomic mass* for an element). The molecular (atomic) mass number is the average composed of the masses (accounting for relative

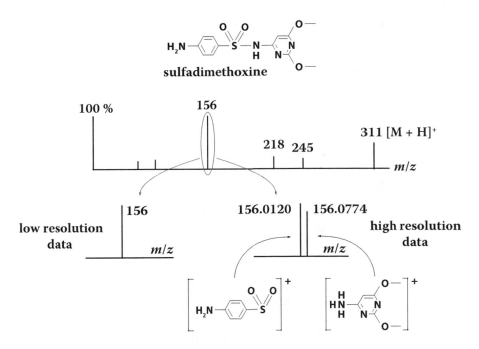

CID spectra of sulfadimethoxine.
High resolution enables the separation of isobaric peaks and
acquisition of additional information such as accurate mass values.
A resolution of ~ 5000 is required to sufficiently separate the isobaric
m/z 156 ions so that accurate mass measurements can be made to
obtain the empirical formulae.

FIGURE 3.5 Separation of isobaric fragment ions at high resolution.

intensities) of all isotopes present in a molecule (element); e.g., among the elements C is 12.01, Cl is 35.5, and Br is 80. The latter two are important examples of the need to distinguish between molecular and monoisotopic mass because inclusion of these elements makes a substantial difference in the calculated mass, i.e., half a mass unit per Cl and one mass unit per Br. As long as the isotopic pattern of a compound is resolved, the peak reported is normally the most intense ion in the cluster, the *monoisotopic ion* (for most small molecules this is the lowest mass in the group). Molecular mass becomes relevant when an observed peak cannot be resolved into its isotopic components, as is typical for proteins. For instance, in the ESI spectrum of apo-myoglobin the peaks are not usually resolved at the isotopic level, which means that each measured *m/z* value is an expression of the molecular mass, and thus the deconvoluted mass of 16,950 is also the molecular mass of this protein (Figure 3.11, shown later).

The exact mass of a compound can be calculated because the accurate atomic masses for the individual elements are known. For example, caffeine is composed of 8 C, 10 H, 4 N, and 2 O atoms, giving it a nominal mass of 194. However, when the exact masses of the elements are added together, the exact mass is 194.0804 Da. Sufficient resolution allows the separation of two compounds that have the same nominal mass but different empirical formulae. As an example, consider if there were the need to differentiate between caffeine ($C_8H_{10}N_4O_2$, 194.0804 Da) and anthracenone ($C_{14}H_{10}O$), which also has a nominal mass of 194 but an exact mass of 194.0732 Da. A resolution of 27,000 would be needed to separate the two compounds at 50% peak height.

The ability to measure the mass of an ion accurately does not necessarily mean that the ion being measured has to be fully resolved from all other possible ions of the same nominal mass. In the case of caffeine and anthracenone, an accurate mass measurement on either ion could be obtained readily on an instrument with a resolution of only 5,000. However, a mixture of the two compounds would not be resolved at 5,000 resolution, and the accurate mass obtained would be the average of the two exact masses. The accurate mass of each can be attained only when the compounds are resolved from each other (Figure 3.6). Note that the peaks for the compounds still overlap at half height at 27,000 resolution; it would take a resolution of ~50,000 to fully separate the two compounds. Therefore, when accurate mass measurements are important, one should buy an instrument with as much resolution as possible (within constraints imposed by the required rate of data acquisition, and available funds).

It is common to use a *mass window* when considering whether the accurate mass obtained reflects a particular empirical formula. A number used frequently is 5 parts per million (5 ppm), representing 0.0001 Da (0.1 mDa) per 20 Da. Five parts per million at *m/z* 100 is approximately the mass of one electron. The usefulness of a ppm mass window depends on where a compound falls on the mass scale because the size

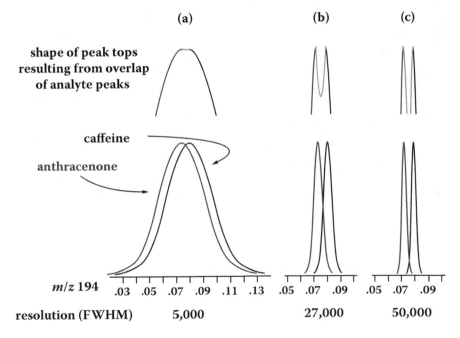

(a) At 5,000 resolution, the accurate masses of anthracenone and caffeine can be measured as individual compounds (they differ by 7 mDa) but a mixture would not be resolved.

(b) At 27,000 resolution, the peaks overlap at 50 % height but accurate mass measurement may still be compromised by the shape of the coalesced peaks.

(c) At 50,000 resolution, the peaks are clearly separated.

FIGURE 3.6   Effect of resolution on accurate mass measurement of mixtures.

of a 5 ppm window increases when expressed in millidalton (or millimass) units. For example, 5 ppm is the equivalent of 1 mDa (1 mmu) at 200 Da, but it is 5 mDa (5 mmu) at 1 kDa. A window the size of which is mass dependent is an inappropriate measure when the objective is to obtain, or confirm, a molecular formula, also because the number of possible empirical formulae increases with mass. An alternative approach is to use a mass window defined by mass instead of ppm. The *Journal of Organic Chemistry* now uses a fixed 3 mDa (mmu) window as the acceptable error on accurate mass measurement for characterization data up to 1 kDa. Although it is not ideal, the move to a fixed window for accurate measurement is preferable to a parameter that varies with mass.

The number of possible empirical formulae increases dramatically even when working with just the common elements of organic and biological chemistry (C, H, N, O, P, S, Si, halogens), depending on the number of elements included. Table 3.2 shows the effects of varying the number of elements and the effect of using either a ppm or mDa (mmu) mass window. It is seen that for the 3 mDa window there is an increase in the number of possible formulae at lower masses, while the number is restricted at higher values. The changeover point is m/z 600, where 3 mDa = 5 ppm. The masses at which the largest numbers of empirical formulae occur change as the mass sufficiencies or deficiencies of the elements are factored into the calculation.

When carbon and hydrogen are the only elements, the mass sufficiency of hydrogen increases the fractional masses of compounds, e.g., adding 0.1252 Da for C_7H_{16} at 100 Da (to yield an accurate mass of 100.1252 Da) and 1.1111 Da for $C_{70}H_{142}$ at 982 Da (accurate mass = 983.1111 Da). The increase in fractional mass is rapidly negated when heteroatoms are included, as shown in Table 3.2. The extent of how much exact masses vary is illustrated by comparing the hydrocarbon $C_{40}H_{82} = 562.6417$ Da, the tripeptide trp-trp-trp (WWW) $C_{33}H_{32}N_6O_4 = 576.2496$ Da, and pentabromobiphenyl $C_{12}H_5Br_5 = 543.6306$ Da ($^{79}Br_5$ isotope). While the nominal masses of these compounds are 562, 576, and 544 Da, respectively, there is a range of more than a full

---

TABLE 3.2

Number of Possible Empirical Formulae at Different Masses

| | Possible Formulae $C_{0-500}H_{3-500}N_{0-15}O_{0-15}$ | | | Possible Formulae $C_{0-500}H_{3-500}N_{0-15}O_{0-15}P_{0-3}$ $S_{0-3}Si_{0-3}F_{0-3}Cl_{0-3}Br_{0-3}$ | |
| | Mass Window | | | Mass Window | |
Mass (Da)	3 mDa	5 ppm	Mass (Da)	3 mDa	5 ppm
100.0500	3	1	100.0000	16	2
250.0750	9	5	249.9250	697	291
500.1500	19	14	499.9000	13,950	11,630
1000.3000	16	28	999.9000	53,764	97,324

Note: The masses with the highest numbers of formulae are different because of the mass deficiencies of P, S, Si, F, Cl, and Br.

mass unit in their mass sufficiencies and deficiencies, i.e., from 0.3694 below unit mass to 0.6417 above unit mass.

It has been argued that determining accurate masses above 1 kDa is of limited value, but it is important to consider that the numbers of possible formulae depend on the sets of components (masses) used to make up the designated maximum mass. There are many possible combinations that can be assembled to reach 1 kDa using the elements C, H, N, O, P, S, Si and the halogens (Table 3.2). However, if only the 20 amino acids are used (with residue masses ranging from 57.0215 Da for glycine to 186.0793 Da for tryptophan; Table 3.8, given later), then the number of possible combinations for a mass of even a few thousand daltons is limited, and the value of accurate mass measurements becomes relevant again.

Despite the discussion above it is incorrect to assign a specific number to the error associated with the accuracy of a measurement, be it of mass or any other measured parameter, because the attainable accuracy of experimentally determined values should be expressed statistically. The current windows of 3 mDa and 5 ppm, pertaining to the accuracy of mass measurements, are compromises that have a historical basis.

The error in accurate mass measurements is composed of two components: the error attributable to the measurement itself, such as the distortion caused by an isobaric interference, and the error arising from instrument design and manufacturing tolerances. Examples include how effectively ions can be focused and the quality of power supplies.

When determining the empirical formula of an unknown, it is important to consider other information, including the isotope pattern of the analyte, the nitrogen rule, the elements present (Table 3.2 illustrates the effect of including extra elements), and other spectroscopic data. Chemically impossible formulae, e.g., N_5H_2, should be ruled out. Care should be taken to ensure that an assigned formula is appropriate; e.g., if a bromine is present, the ^{79}Br and ^{81}Br isotopes must occur in a 1:1 ratio. Experience working with accurate mass values can give an indication of the nature of a compound, particularly when C, H, N, and O are the only elements involved, as hydrogen is the most frequently occurring element with a fractional mass above unit, i.e., 1.007825 Da. Saturated compounds contain a large number of hydrogens, thus their exact masses are considerably higher than their nominal mass, e.g., the exact mass of the alkane $C_{14}H_{30}$ is 210.2348 Da. In contrast, aromatic compounds have far fewer hydrogen atoms, so their exact masses are much closer to their nominal masses; e.g., the exact mass of pyrene ($C_{16}H_{10}$) is 202.0783 Da. Furthermore, the exact masses of all elements heavier than nitrogen are less than unit (they are mass deficient), thus their presence can make substantial differences in exact mass values; e.g., iodobenzene, C_6H_5I, has an exact mass of 203.9436 Da, whereas the exact mass of benzene, C_6H_6, is 78.0470 Da. The mass *deficiency* of iodine counteracts the mass *sufficiency* of hydrogen (1 iodine counteracts >12 hydrogens).

When multiple measurements are made of a parameter, such as mass, the precision of the result should be determined. *Precision* is a measure of how closely clustered are the values of a set of repeat measurements, but it does not provide information as to whether the data are accurate. For example, if the calibration of the mass scale is incorrect, then replicate mass measurements may be very close to each other, i.e., they are very precise, but the results will not be accurate because of the error caused by the incorrect calibration. While precision is a valuable asset, it should not be confused with accuracy, as it possible for a set of data to be "precisely wrong."

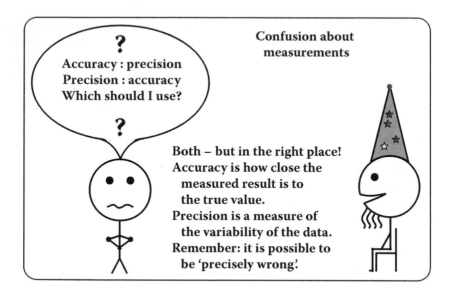

The term *high-resolution mass spectrometry* is a misnomer, often associated with accurate mass measurement. This moniker has long been used as a substitute for accurate mass measurement. The fact that an instrument can be operated at high resolution does not necessarily mean that accurate mass values are guaranteed. Consider the case of caffeine and anthracenone illustrated in Figure 3.6: high resolution (27,000) is necessary to separate the molecular ions of these two compounds, but attaining this separation does not automatically provide the accurate mass of either compound. The ability to obtain accurate mass measurements depends on the calibration of the mass scale of the analyzer. Furthermore, the accurate mass on either compound (but not in a mixture) can be obtained at a resolution of 5,000. Of course, high resolution is an important factor for obtaining accurate mass measurements because it separates isobaric species and facilitates precise (closely clustered) measurements.

3.2 INTERPRETATION OF MASS SPECTRA

The amount of energy acquired during ionization and the structure of the analyte determine the complexity of the mass spectra obtained. Other features, including isotopic composition, the nitrogen rule, rings plus double bond considerations (see later), and accurate mass values, provide additional information.

3.2.1 WHAT IS A MASS SPECTRUM?

As noted previously, a *mass spectrum* is a graphical representation of the *m/z* values (x-axis) of the ions together with their intensities (y-axis), collected over a specified mass range (Figure 1.3). The x-axis represents the mass-to-charge ratio (m/z), where m is the total of the masses of the component atoms and z is the number of charges on the ion (positive or negative) expressed in whole numbers (1, 2, 3, etc.). A peak on the x-axis occurs where an increase in the number of ions is detected and a signal pertaining to a specific m/z is registered. The y-axis represents the relative intensity of the ions, usually expressed with respect to the most stable, and therefore most intense (100%) ion in the spectrum, the *base peak*. It is also possible to use absolute values on the y-axis, e.g., ion current, but this is rarely done. The recognition and importance of multiply charged ions is discussed in Section 3.2.4.

Various ion types may be observed in the molecular mass region. If the analyte remains *intact* upon ionization (in positive mode), it is observed either as the molecular ion, $[M]^+$, when using EI, or as an ion formed by the addition of a charged species to the neutral molecule when other ionization methods are utilized. The charge is often provided by the addition of a proton to the analyte to form a protonated molecule, $[M + H]^+$. Numerous other molecular adducts can be formed depending on the ionization method, e.g., $[M + Na]^+$ and $[M + K]^+$ in ESI, while $[M + C_2H_5]^+$ and $[M + NH_4]^+$ occur in CI (depending on the reagent gas). In negative ionization, which is used less frequently, the ions generated include those formed by the acquisition of an electron, $[M]^-$, the loss of a proton, $[M - H]^-$, or the addition of a halogen or other negative ion, e.g., $[M + Cl]^-$ and $[M + CH_3COO]^-$. Table 3.3 shows ions observed in the molecular ion region for several important ionization techniques. Another reminder: The term *protonated molecular ion* is incorrect because such an ion would be doubly charged.

TABLE 3.3
Typical Ions Observed in the Molecular Ion Region

Source Type	Positive Ions (cations)	Negative Ions (anions)	Comments
EI	$[M]^+$		
ECNI		$[M]^-$	Methane/nitrogen as buffer gas
CI	$[M + H]^+$		
	$[M + CH_5]^+$		Methane reagent gas
	$[M + C_4H_9]^+$		Isobutane reagent gas
	$[M + NH_4]^+$		Ammonia reagent gas
APCI	$[M + H]^+$	$[M - H]^-$	
APPI	$[M]^+$		
	$[M + H]^+$		Mediated via a dopant
DESI	$[M + H]^+$	$[M - H]^-$	
DART	$[M + H]^+$	$[M + O]^-$	Mediated via H_2O (+) and O_2 (−)
MALDI	$[M + H]^+$	$[M - H]^-$	
	$[M + Na]^+$		
ESI	$[M + H]^+$	$[M - H]^-$	
	$[M + \text{alkali metal}]^+$	$[M + \text{halogen}]^-$	
	$[M]^+$	$[M]^-$	Organometallics
	$[M + nH]^{n+}$		Peptides/proteins

3.2.2 Isotope Patterns

There is an obvious impact on the *isotope pattern* of the spectrum when an analyte includes certain heteroatoms (Figure 3.7). Among commonly encountered elements displaying distinctive isotope patterns are chlorine, where the ratio of the ^{35}Cl and ^{37}Cl isotopes is 3:1, and bromine, where the ratio of ^{79}Br to ^{81}Br is 1:1. There is a 2 Da separation between the isotopes for both these elements. When there are multiple copies of an element, the isotope frequencies are obtained by cross-multiplying the ratios of the isotopes for each atom present, e.g., for two chlorines $(3:1)(3:1) = (3 \times 3):(3 \times 1 + 1 \times 3):(1 \times 1) = 9:6:1$. For three chlorines the ratio is 10:9:3:0.3, while two and three bromines give ratios of 1:2:1 and 1:3:3:1 ratios, respectively (Figure 3.7).

Many metals give distinctive isotope patterns, e.g., tin, ruthenium, osmium, platinum, palladium, and others. The first two are shown in Figure 3.7. The isotope frequencies for a few major elements are given in Table 3.4. Certain uncommon

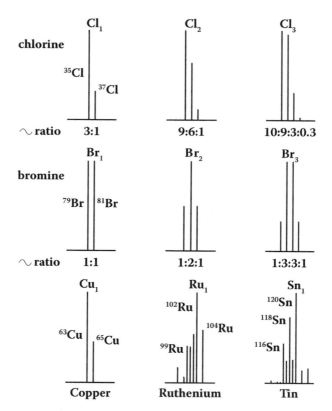

Certain elements affect the isotope patterns of spectra dramatically. Illustrations include: 1 – 3 chlorine and bromine atoms; copper (similar to chlorine); ruthenium and tin (both have several isotopes >10%).

FIGURE 3.7   Examples of isotope patterns.

TABLE 3.4
Nuclidic Masses and Relative Abundances for Naturally Occurring Isotopes of Some Important Elements

Element	Symbol	Mass	Relative Intensity
Hydrogen	^{1}H	1.007825	100
(Deuterium)	$D\ (^{2}H)$	2.0141	0.015
Carbon	^{12}C	12.0000	100
	^{13}C	13.0034	1.1
Nitrogen	^{14}N	14.0031	100
	^{15}N	15.0001	0.4
Oxygen	^{16}O	15.9949	100
	^{18}O	17.9992	0.2
Sodium	^{23}Na	22.9898	100
Silicon	^{28}Si	27.9769	100
	^{29}Si	28.9765	5.1
	^{30}Si	29.9738	3.4
Phosphorus	^{31}P	30.9738	100
Sulfur	^{32}S	31.9721	100
	^{34}S	33.9679	4.5
Chlorine	^{35}Cl	34.9689	100
	^{37}Cl	36.9659	32.0
Potassium	^{39}K	38.9637	100
	^{41}K	40.9624	7.2
Bromine	^{79}Br	78.9183	100
	^{81}Br	80.9163	97.3
Tin	^{112}Sn	111.9048	3.0
	^{114}Sn	113.9028	2.0
	^{115}Sn	114.9033	1.0
	^{116}Sn	115.9017	43.6
	^{117}Sn	116.9030	23.6
	^{118}Sn	117.9016	73.3
	^{119}Sn	118.9033	26.4
	^{120}Sn	119.9022	100
	^{122}Sn	121.9034	14.2
	^{124}Sn	123.9053	17.8
Iodine	^{127}I	126.9045	100
Platinum	^{192}Pt	191.9610	2.3
	^{194}Pt	193.9627	97.4
	^{195}Pt	194.9648	100
	^{196}Pt	195.9649	74.6
	^{198}Pt	197.9679	21.2

Note: Tin and platinum are examples of metals with many isotopes. Some elements such as phosphorous and iodine have only one isotope.

elements can cause confusion; e.g., silver has a 1:1 ratio of masses 105 and 107 and looks like bromine, while copper has an isotope pattern similar to that of chlorine (Figure 3.7). Yet other elements are distinctive because they have an isotope one mass lower than their dominant isotopes; notable are the light elements, lithium and boron, where ^6Li (8%) and ^{10}B (25%) are one mass lower than the primary isotopes, ^7Li and ^{11}B. There are other elements with subtle isotope patterns that are still useful in indicating their occurrence, e.g., sulfur, ^{34}S (4%), and silicon, ^{29}Si (5%) and ^{30}Si (3%). The presence of these elements is suspected when the M + 2 ion is more intense than predicted by the frequency of two ^{13}C atoms.

An important use of isotope profiles, particularly when C, H, N, and O are the only elements present, is to establish the number of carbon atoms in a molecule. The natural abundance of ^{13}C is 1.1% that of ^{12}C; i.e., there is a 1.1% chance that any given carbon in a molecule will be a ^{13}C, thus the number of carbon atoms can be determined from the intensity of the M + 1 ion. For instance, an intensity of 22% for ^{13}C indicates that there are 20 carbon atoms in that species. The frequencies of ^2H, ^{15}N, and ^{18}O are 0.015, 0.37, and 0.2%, respectively, so the effect of ^2H is minimal, and even if there were two nitrogen atoms in a molecule, the ^{15}N would only add 0.75% to the intensity of M + 1; also, any ^{18}O would be irrelevant as this isotope would be seen as part of the M + 2 ion. Other examples of the effect of ^{13}C contribution are illustrated in Figure 3.8. Once the number of carbon atoms increases to C_{91}, it becomes more likely than not that there

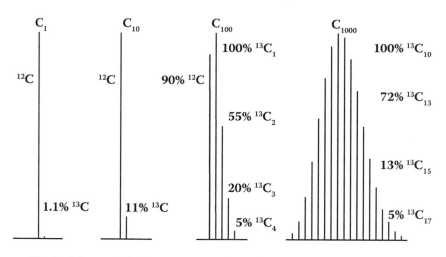

The incidence of ^{13}C is 1.1% per carbon, thus, as the number of ^{12}C atoms increases so does the intensity of the ^{13}C peak.
A point is reached where the intensity of the 'carbon-13' peak exceeds that of the 'carbon-12' peak. The figure shows C_{100} where the $^{12}C_{99}{}^{13}C_1$ ion is more common than the $^{12}C_{100}$ ion.
For C_{1000} the most common ion will include ten ^{13}C atoms ($^{12}C_{990}{}^{13}C_{10}$) and the frequency of the $^{12}C_{1000}$ ion is 0.01%.

FIGURE 3.8 Carbon isotope distribution.

will be at least one ^{13}C somewhere in the molecule; thus, the peak containing one ^{13}C will be more intense than the one where only ^{12}C is present. As illustrated in the figure, by the time C_{1000} is reached, the most intense ion of the isotopic cluster will contain 10 ^{13}C atoms (i.e., $^{12}C_{990}{}^{13}C_{10}$) and the intensity of the $^{12}C_{1000}$ ion will be minimal (abundance = 0.01%). The molecular (average) mass of a compound containing C_{1000} will include 10.7 ^{13}C atoms. Cases where the number of carbon atoms causes the carbon-13 ion to be more intense than the carbon-12 ion are uncommon in EI spectra, but are regularly seen in the ESI and MALDI spectra of biomolecules.

3.2.3 ELECTRON IONIZATION SPECTRA OF SMALL MOLECULES

Electron ionization (EI) spectra are likely to be complex because they almost always contain fragment ions resulting from the energy imparted by the bombarding electrons that are accelerated to 70 eV, significantly above that required to ionize organic compounds (~10 eV). EI remains a major ionization method for small nonpolar compounds analyzed by GC-MS.

EI spectra often include many fragment ions; there are long-standing rules for the interpretation of the fragmentation processes. Likely losses for known compounds can be determined by looking at the structure of the compound and considering which small groups could readily fragment, e.g., methyl and methoxy groups. The first loss is usually a fragment with an odd mass while subsequent losses are often even masses. For example, in alkanes the first loss is a $-CH_3$ (15 Da), and all subsequent fragments reflect the progressing removal of $-CH_2$ (14 Da) groups. The locations of the heteroatoms are important with known structures; fragmentation is more likely at such points, as the link between these atoms and their adjacent carbons is generally weaker than that of carbon:carbon linkages. Table 3.5 provides a list of common losses from molecular ions. When the compound is an unknown, the reverse inferences can be made to provide structural information, e.g., loss of 31 Da from the molecular ion implies the presence of a methoxy ($-OCH_3$) group.

When interpreting EI spectra, one should first find the ion with the highest m/z (this may be an ion of very low intensity, <1%) and then decide if it is the molecular ion. The easiest way to do this is to consider if the ion with the next highest m/z value can be interpreted as a fragment of the purported molecular ion. If this fragment ion is more than 2 units (H_2) and less than 15 units ($-CH_3$) below the m/z of the initially chosen ion, then that ion cannot be the molecular ion because there is no logical fragment that could be removed to attain the lower m/z value. Also, there is no chemically sensible fragment that can be used to explain apparent losses of 21–25 Da. In either case, the molecular ion must have had a higher m/z value and undergone fragmentation to yield the observed ion. Next, one should make certain that the suggested molecular mass and fragments are in agreement with the following rules and considerations: strong molecular ions indicate aromaticity, multiple ions separated by 14 Da ($-CH_2-$) indicate a hydrocarbon, and subgroups such as $-CH_3$, $-C_2H_5$, $-CH_3O$, $-CH_3CO$, and $-C_4H_9$ yield straightforward (odd mass) losses from the molecular ions of 15, 29, 31, 43, and 57 Da, respectively. Furthermore, the isotope pattern of the spectrum can suggest the inclusion of certain elements, such as halogens or metals (Figure 3.7), and the number of carbon atoms can be calculated from the intensity of the ^{13}C isotope peak (Figure 3.8).

TABLE 3.5

Common Losses from Molecular Ions Observed in EI

Mass Lost	Group
1	H
2	H_2
15	$-CH_3$
17	$-OH$
18	H_2O
20	HF
26	C_2H_2
27	HCN
28	C_2H_4, CO, N_2
29	$-C_2H_5$, $-CHO$
31	CH_3O-
32	CH_3OH
36	HCl
43	$-C_3H_7$, CH_3CO-
44	CO_2
57	$-C_4H_9$
73	$-Si(CH_3)_3$
77	$-C_6H_5$

Another consideration is the *nitrogen rule*, which states that if a neutral species containing the elements common to organic molecules (C, H, N, O, S, Si, P, halogens) has an odd molecular mass, then there must be an odd number of nitrogen atoms present. Conversely, an even molecular mass indicates the presence of zero or an even number of nitrogens. The nitrogen rule is a mathematical consequence of the fact that nitrogen (^{14}N) is the only element with an even atomic mass that has an odd valence.

Double bond equivalents (DBEs) are a tool that reflects the degree of saturation of an unknown by giving the number of rings and double bonds. $DBE = m(C + Si) - n(H + halogen)/2 + o(N + P)/2 + 1$, where m, n, o = the number of atoms in each group, e.g., benzene has a DBE of 4 (one ring and three double bonds). Whether ions have odd or even masses and carry odd or even numbers of electrons provides useful information but can be confusing. Molecules (except stable radicals) have an even number of electrons; therefore, an odd electron species is created when an electron is removed in EI. Such an entity with an odd electron count is called a *radical ion*. As noted earlier, a radical molecular ion should be written $[M]^{+\bullet}$ to signify that it is a radical with both an electrical charge and an unpaired electron; however, in mass spectrometry, one is more likely to see $[M]^+$ written. Cleavage of a molecular ion tends to remove a neutral radical (an odd electron species), creating an ion with an even number of electrons. Once such an ion is formed, further losses are even electron species; thus, the resultant fragments will have an even number of electrons. These ions will be observed at odd *m/z* values, except when there are odd numbers of nitrogens when the even electron species occur at even *m/z* (Figure 3.9).

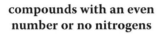

compounds with an even number or no nitrogens

e.g., C_mH_n

$M^{+\cdot}$ odd electron, even mass, radical

$-CH_3$ ↓

F1   even electron, odd mass

$-CH_2$ ↓

F2   even electron, odd mass

$-CH_2$ ↓

F3   even electron, odd mass

compounds with odd number of nitrogens

e.g., C_mH_nN

$M^{+\cdot}$ odd electron, odd mass, radical

$-CH_3$ ↓

F1   even electron, even mass

$-CH_2$ ↓

F2   even electron, even mass

$-CH_2$ ↓

F3   even electron, even mass

In both cases, electron ionization results in the formation of an odd electron radical.

The first loss is usually an odd electron species, e.g. CH_3, yielding an even electron fragment.

Subsequent losses (e.g. CH_2) are of even electron species yielding even electron fragments.

Whether the mass of a molecular ion and its fragments are odd or even depends on the nitrogen rule.

FIGURE 3.9   Odd:even electron species and their masses.

Odd vs. Even

Odd electron have even masses; even electron have odd masses: but if odd numbers of nitrogens are present then even electron ions have even masses!

This all makes *perfect* sense to me!

Of course! Just think of the classic "Who's on first" baseball sketch or the "Rules of cricket" poem.

3.2.4 SPECTRA FROM IONIZATION BY THE ADDITION OR REMOVAL OF CHARGES

Mass spectra generated by ionization methods that form ions by the addition or removal of charged species (CI, APCI, ESI, MALDI, i.e., the soft ionization methods) are much simpler than EI spectra, because these methods impart less energy to the analyte than EI, resulting in minimal or no fragmentation. The molecular species usually constitute the base peak, e.g., $[M + H]^+$, $[M + NH_4]^+$, $[M + Na]^+$, $[M + K]^+$, $[M – H]^-$, $[M + Cl]^-$. Such spectra are straightforward and valuable, as they provide the molecular mass of the analyte. However, no structural information is obtained beyond that available from the isotope pattern, the nitrogen rule, and (when available) accurate mass data. All molecular e.g., $[M + H]^+$, $[M + Na]^+$, etc., are even electron species and not radicals, as all electrons are paired.

3.2.4.1 Electrospray Ionization Spectra of Large Molecules

The structures of large molecules, including all biopolymers, have multiple sites where charges can be added during ESI, leading to the formation of envelopes of ions with *multiple charge states* that are observed at their respective *m/z* values. For instruments with sufficient resolution (i.e., analyzers other than quadrupoles), the spacing between the isotopes can be used to determine the charge state of the ions for compounds with molecular masses up to a few thousand daltons (higher for FT-ICRMS instruments). The isotope spacings observed are 1 divided by the charge state of an ion because isotopes are, by definition, one mass unit (neutron) apart (e.g., ^{12}C and ^{13}C), and the x-axes of spectra are *m/z*. The spacing on the *m/z* axis is 1 divided by 2, 3, and 4, respectively, for doubly, triply, and quadruply charged ions; thus, the apparent spacings are 0.5, 0.33, and 0.25 mass units (Figure 3.10).

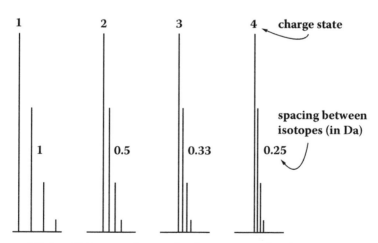

**With sufficient resolution the charge state of an ion can
readily be determined from the spacing of the isotopes.
For instance, spacing for a singly charged ion is 1 Da
while for a quadruply charged ion it is 0.25 Da.**

FIGURE 3.10   Charge state vs. isotope spacing.

When a peptide or protein is analyzed, one should look not only for the singly charged ion, as it may not be present, but also for ions representing multiple charge states. For example, a peptide of 2,000 Da may not show an ion at m/z 2,001 corresponding to $[M + H]^+$, but will likely exhibit ions at $[M + 2H]^{2+}$, $[M + 3H]^{3+}$, and $[M + 4H]^{4+}$ at m/z 1,001, 667.7, and 501, respectively. All these ions are molecular species and are equally valid representations of the peptide. Just because the $[M + H]^+$ is not observed, it is incorrect to think that the expected compound is not present. Ions from proteins exhibit many charge states, e.g., the spectrum of horse heart myoglobin (apo form) shows ions in the +20 to +11 charge state range, between m/z 848 and 1,542 (Figure 3.11a). Notice the progressively increasing spaces between the peaks; such spacings are always present for biopolymers.

The progressive spacing of ions in the ESI mass spectra of biopolymers is a mathematical consequence of dividing a given (single) number by a consecutive set of digits. Thus, for a 10 kDa protein with charge states of 15, 14, 13, 12, 11, and 10, ions will occur at m/z 668, 715, 770, 834, 910, and 1,001, respectively, when the masses of the added protons are included. Therefore, the spacing between the ions will be

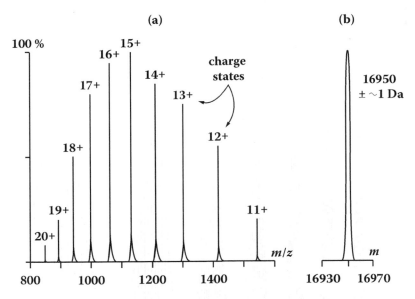

(a) The progressively increasing spacing between the
    multiply charged ions indicates a biopolymer, most often
    a protein.
(b) Deconvolution (transform) of the multiple charge states
    provides the molecular mass. This is often referred to as
    the 'zero charge' state. The deconvolution software
    solves a set of simultaneous equations derived from
    the spacing of the charge states.

FIGURE 3.11 Electrospray mass spectrum of apo-myoglobin and the deconvoluted data giving the molecular mass of the protein.

47, 55, 64, 76, and 91 Da, i.e., a set of numbers where the differences between values form an increasing series. The intensity of the distribution will appear *skewed* to lower *m/z* values (as seen in Figure 3.11). Furthermore, the isotope envelope at each *m/z* will become narrower because it is divided by the charge (Figure 3.12). The solution of a set of simultaneous equations derived from the differences between the charge states, called *deconvolution*, yields the molecular mass of the protein or other biopolymer. Computer software is readily available for such calculations.

The masses of the protons added during ionization to produce each multiple charge state (e.g., 15 protons for the $[M + 15H]^{15+}$ ion) are removed during deconvolution thus giving the neutral, zero charge state of the protein (Figure 3.11b). The mass obtained is the molecular mass into which all the isotopic species have been averaged, reflecting the fact that the isotopes are not resolved from each other for the multiply charged ions.

Accurate calibration of the mass scale is especially important when determining the masses of biopolymers because errors are magnified during deconvolution, e.g., a 0.1 Da error for a 10 charge state ion becomes a 1 Da error when the observed *m/z* value is multiplied by 10 to obtain the mass of the analyte. With good mass calibration, the experimental error in the molecular mass for a small protein, such as myoglobin, is ~1 Da.

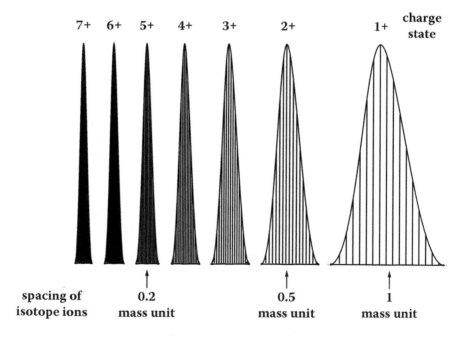

The envelopes become narrower as the charge state of the ions increases because the spacing between isotopes is divided by the number of charges on the ion.

FIGURE 3.12   Effect of charge state on the width of the isotope envelope.

The characteristics of the ESI mass spectra of biopolymers are in contrast to those observed for the spectra of synthetic polymers, such as polyethylene glycol (PEG) (Figure 3.13). Each m/z value corresponds to a single compound for these polymers. In the case of PEG with a low average molecular mass, e.g., ~500 Da, each separate

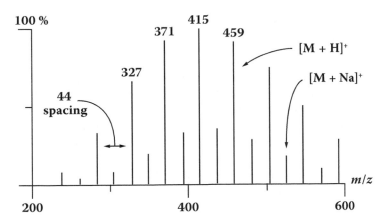

Even spacing of ions that is characteristic for synthetic polymers occurs because each ion corresponds to a separate component of the polymer.
For PEG the space between the singly charged major ions, m/z 44, corresponds to the $-CH_2CH_2O-$ repeat group.

FIGURE 3.13   Electrospray spectrum of polyethylene glycol (PEG).

component carries a single charge, yielding ions that are spaced *evenly*, 44 Da apart, corresponding to the $-CH_2CH_2O-$ repeat unit. Note that doubly or triply charged envelopes may also be observed for PEG with higher average molecular masses, but the spacing between the individual components of each envelope will still be constant (and consistent), i.e., $44/2 = 22$ and $44/3 = 14.7$. The spectra of synthetic polymers frequently (but not always) display Gaussian distributions rather than the skewed patterns seen for biopolymers.

3.2.4.2   Atmospheric Pressure Ionization and Fragmentation

The API methods, which have vastly extended the range of mass spectrometry, produce simple spectra that often exhibit only the molecular mass of the analyte with no (or minimal) fragmentation, and little structural information. Methods were required to induce fragmentation and thereby increase the information content of API spectra. The techniques developed, often tailored to specific objectives, control the amount of energy added to the analytes so that the fragmentation is directed to certain chemical bonds. This is particularly relevant for the analysis of biopolymers where the fragmentation takes place at specific points in the structure, e.g., the peptide bond, and provides sequence and other structural information.

A distinction is made between fragmentation inside the ion source and those occurring in other locations. Although the processes taking place in MS/MS are still fragmentations, specific names are used to indicate the type of processes, e.g., *collision-induced dissociation* (Section 3.3.2.1), *infrared multiphoton dissociation* (Section 3.3.2.2), *electron transfer dissociation and electron capture dissociation* (Section 3.3.2.3), as well as the more specialized *photodissociation* and *surface-induced dissociation* (Section 3.3.2.4).

3.2.5   FRAGMENTATION AND QUANTIFICATION

The intensity of the current produced by analyte ions is relevant in quantification. Limits of detection are improved when fragmentation is reduced or eliminated and the ion current, attributable to the analyte, is present as a single species. For instance, using CI often improves both detection and quantification limits when compared to EI, although the controlled fragmentation used in selected reaction monitoring can also improve detection limits. Fragmentation as it applies to specific quantification techniques for small molecules is discussed in connection with the quadrupole family of instruments (Sections 3.3.3.1 and 3.3.5). Quantification for biopolymers, particularly proteins, is presented in Section 3.5.1.9.

3.2.6   A FEW IONS TO AVOID (OR AT LEAST RECOGNIZE)

Despite best efforts, ions are encountered that do not appear to have any relationship to the analyte. Such ions are usually due to contaminants. While extensive lists of common contaminants and their respective interfering ions are available, e.g., on instrument manufacturers' websites, some merit discussion.

The phthalates impart flexibility to plastics, particularly to Tygon tubing. A common member of this family is dioctylphthalate, usually di-(ethyl-hexyl)phthalate, the

EI spectrum of which is dominated by m/z 149 (base peak) and has a molecular ion at m/z 390 of low intensity (<1%). In ESI, [M + H]$^+$ at m/z 391 and [M + Na]$^+$ at m/z 413 are observed. Various dioctylphthalate spectra, obtained using different ionization methods, are shown in Figure 2.8. A benefit of the ubiquitous presence of dioctylphthalate is that the m/z 391 or m/z 413 ion can be used as a lock mass for accurate mass measurements in ESI-TOF instruments in flow injection analysis.

Contaminants from GC columns occur with increasing intensity as the oven temperature is increased. For the commonly used liquid phases (DB-5, Rtx-5, HP-5, etc., with 95% methyl:5% phenyl substituent groups), column-derived ions are observed at m/z 207, 281, and 355. They are present as a continuous bleed that increases in intensity as the column temperature is raised. The separation between these ions corresponds to the dimethyl siloxane subunits ($-(CH_3)_2SiO-$, 74 Da). Septa in injection ports and the seals on sample vials also contribute siliconates, often in the m/z 400–600 range; these compounds are observed as distinct peaks eluting from the GC. A major giveaway of siliconates is their isotope patterns because the ^{29}Si and ^{30}Si isotopes of the multiple silicon atoms contribute to [M + 1]$^+$ and [M + 2]$^+$ (and higher) ions that are much more intense than would be expected for compounds containing only C, H, N, or O.

ESI spectra often contain an envelope of ions separated by 44 Da, originating from polyethylene glycol (PEG). The repeat unit is the ethylene glycol residue, $-C_2H_4O-$ (44 Da). Sources include the ethylene glycol-based detergents and TRIS buffer with TWEEN, as well as the polyethylene glycols themselves. Ions separated by 58 Da correspond to the propylene-glycol repeat unit ($-C_3H_6O-$).

In ESI spectra the lack of a ^{13}C ion indicates the presence of inorganic salts. The spectra of these compounds often show peaks that repeat up to several hundred m/z, resulting from the formation of clusters where the ions are of the form [salt$_n$ + charge]$^{+/-}$. Examples include sodium trifluoroacetate with a repeat unit of 136 (CF_3COONa) that gives ions corresponding to [($CF_3COONa)_n$ + Na]$^+$ in the positive ion mode and [($CF_3COONa)_n$ + CF_3COO]$^-$ in the negative ion mode. In the case of sodium chloride the repeat is 58 (NaCl), and in this case there are the obvious isotope patterns attributable to the different numbers of chlorine atoms.

Certain combinations of inorganic salts give intensity maxima in spectra, the "magic number" ions. For instance, for NaCl there are intense ions corresponding to [(NaCl)$_{10}$ + Na]$^+$ and [(NaCl)$_{13}$ + Na]$^+$ at 606.6 and 782.4 (major isotope), respectively. Although inorganic clusters are often annoying, they can be used to advantage, e.g., sodium formate, and mixtures of cesium, rubidium, and sodium iodides can be used as ESI calibration standards in both positive and negative ion modes.

3.3 ANALYTICAL TECHNIQUES AND STRATEGIES

Mass spectrometric systems come in a variety of configurations: multiple sample introduction and ionization methods may be combined with single analyzers or coupled into MS/MS formats, using the same or different types of mass analyzer. Although most instrument formats are capable of multiple functions, one configuration is often better suited to a specific analysis. For instance, the collection of spectra in TOF instruments is much faster and more sensitive than with quadrupoles; therefore, TOF systems are appropriate for collecting full spectra when monitoring high-speed processes, e.g., explosions. There are also situations where particular analyses are only possible with one type of system, e.g., the isotopic resolution of an FT-ICRMS.

3.3.1 Techniques for Instruments with Single Analyzers

Despite the current emphasis on MS/MS instrumentation for a rapidly increasing number of applications, single-analyzer (MS) systems remain important because they provide valuable information for identification of compounds or the monitoring of selected ions for quantification. Single analyzer instruments are usually easier to operate and maintain, and less expensive, than MS/MS systems.

All quadrupole, time-of-flight, magnetic, and FT analyzers, combined with almost all types of ion source, including EI, CI, API, and MALDI, can be used in single-analyzer formats. The various common sample introduction methods (GC, LC, batch inlet) are also applicable. Structural information, accurate mass measurement of molecular ions (and fragments), and selected ion monitoring for quantification are available, depending on the type of analyzer used. Table 3.6 lists some applications for different types of single-analyzer instrument, GC-MS, LC-MS, and MALDI-TOF.

3.3.1.1 Techniques for Ion Activation

Aside from the conventional generation of fragments (in the ion source) during ionization, there are several other means of providing additional energy to induce fragmentation between the source and the analyzer or inside the analyzer. The type and quantity of fragments are influenced by the amount of energy acquired during the activation of the ions and the distribution of that energy. Simple bond cleavages are favored when small amounts of internal energy are accumulated, while at higher internal energies there is increased fragmentation that is information rich but often difficult to interpret.

TABLE 3.6
Examples of Applications for Single-Analyzer Instruments

Application	Instrument	Data Acquired
Identification		
All fields	TOF LC-MS (API)	Accurate mass measurement for empirical formula determination
Environmental	Quadrupole GC-MS (EI)	Spectral matching to libraries
Synthetic chemistry	TOF-MS (API) flow injection analysis	Accurate mass measurement for empirical formula determination
Biology	TOF-MS (ESI)	Intact protein molecular mass determination
Biology	MALDI-TOF	Peptide mass maps
		Intact protein molecular mass determination
Quantification		
Environmental	Quadrupole GC-MS (CI)	SIM for quantification
Pharmaceutical	Quadrupole LC-MS (API)	SIM for quantification

3.3.1.2 Ion Activation and Fragment Formation in Single-Analyzer Instruments

When an analyte is ionized, the molecular ion formed either remains intact or decomposes immediately (lifetime $<10^{-7}$ s) into fragment ions while still in the ion source. Most ions, molecular or fragment, leaving the ion source travel unchanged through the mass analyzer and are detected by the ion collector.

Some ions may undergo *unimolecular decomposition* (on a microsecond to millisecond timescale) as they leave the source and travel through the analyzer. Ions undergoing such transitions are called *metastable* ions. This type of ion used to have importance in magnetic type analyzers; the only area where they are currently of minor relevance is in MALDI-TOF instruments.

3.3.1.2.1 Fragmentation in MALDI-TOF Instruments

Although MALDI is a soft ionization technique that produces mostly protonated molecules, some *in-source* fragmentation does occur. The resulting ions are observed in both linear and reflectron modes without altering operational parameters. The energy acquired during ionization and acceleration into the flight tube may also be sufficient to cause the formation of metastable species that fragment as they travel through the field-free region of the analyzer prior to detection. Such fragmentation is called *post-source decay* (PSD). When a MALDI-TOF is used in *linear* mode, there is no evidence of PSD because the conservation of momentum dictates that the fragments arrive at the detector as though the initial protonated molecule had remained intact. Alternatively, the precursor and PSD ions can be separated according to the differences in their kinetic energies when a *reflectron* (Section 2.3.3) is present. To obtain PSD data, multiple spectra must be collected using different reflector voltage settings to cover the necessary mass ranges with concatenation of the data to form a single spectrum that shows the fragments obtained over the entire mass range. PSD and in-source decay techniques have now been largely superseded by CID type experiments carried out on TOF/TOF instruments (Sections 2.3.7.1.2 and 3.3.3.2).

3.3.1.2.2 In-Source Fragmentation in API Sources

An important strategy to obtain fragmentation in single-analyzer instruments (with API sources) is to increase the voltage at the point where ions enter the vacuum chamber. Significant fragmentation of the analytes often occurs at higher voltages, so that both protonated molecules and intense fragment ions are formed. Another consequence of providing additional energy (when using ESI) is a reduction in the charge state of multiply charged species. Two important advantages of in-source fragmentation are the possibilities of obtaining MS/MS-like spectra without a tandem analyzer, and providing intense fragments for subsequent second-generation *collision-induced dissociation* (CID) ions (MS^3) in tandem instruments. Other advantages include simple operation (often programmable), high sensitivity, and no loss of resolution. However, the technique is much less specific than CID (Section 3.3.2.1) because for mixtures of analytes all ions are fragmented at the same time, and it is usually not possible to relate the fragments to specific precursor ions.

3.3.2 Mass Spectrometry/Mass Spectrometry (MS/MS)

When there is more than one analyzer in an instrument, the parameters of the individual analyzers can be varied, separately or in combination, to obtain full spectra or to monitor selected ions. The strategy of using two analyzers in combination is called *mass spectrometry/mass spectrometry* (MS/MS).

The fundamental process of MS/MS (in a tandem mass spectrometer) is that a particular *precursor* ion is fragmented (decomposed) into a smaller *product* ion accompanied by the loss of a neutral fragment. The precursor ion is usually selected in the first analyzer, reacted by some means in a specialized chamber (see below) to produce product ions that are then separated in a second analyzer, hence the terms *mass spectrometry/mass spectrometry* and *tandem mass spectrometry*. As discussed later, precursor and product ion analyses can be *separated in space*, using sequential analyzers that are either of the same type or in hybrid configurations, or *separated in time*; in the latter the two analytical processes are carried out sequentially, within the same analyzer. There are several MS/MS instrument combinations, including QqQ, QIT, LIT, TOF/TOF, QTOF, LIT-orbitrap, and LIT-FT-ICRMS.

Changes have been made concerning the names of the ions generated by MS/MS. Ions originally named *parent* ions are now called *precursor* ions, and *daughter* ions are now known as *product* ions. *Granddaughter* ions are now *third-generation* ions, and in some ion trap analyses there are *fourth-generation* and higher ions. Although the earlier, familial, names are no longer acceptable, they are still heard frequently.

3.3.2.1 Collision-Induced Dissociation (CID)

Before discussing MS/MS techniques, it is important to consider the processes that occur in a specialized chamber, called a *collision cell*, placed between the analyzers. In this cell the analyte ions, to which translational energy has been added during their acceleration into the cell, collide with neutral atoms with a high ionization potential (usually argon) or molecules (e.g., nitrogen), maintained at relatively high pressure ($\sim$6 mTorr, 8×10^{-3} mbar). During these collisions a portion of the ions' kinetic energy is converted into excess internal vibrational or electronic energy; the process is called *collisional activation*. If the excess energy acquired during the collisions is adequate to break chemical bonds, then these ions are said to have undergone *collision-induced dissociation* (CID), also called *collisionally activated dissociation/decomposition* (CAD). While CID is the most common process that takes place in collision cells, there are other techniques that may utilize the space between the first and second analyzers, e.g., electron transfer dissociation (ETD) and ion mobility separation (IMS).

The CID process is cumulative. The ions enter the CID cell with kinetic energy acquired during their acceleration (10 to 100 V) from the first analyzer. Once in the cell, the ions accumulate additional energy incrementally through multiple collisions with the neutral gas atoms until they fragment into a (limited) variety of product ions (Figure 3.14). Fragmentation normally involves breaking the bond between a carbon and an adjacent heteroatom, e.g., N, O, S, and Si. The nature and degree of fragmentation depend strongly on the initial accelerating voltage and, to a lesser degree, on the type and pressure of the collision gas.

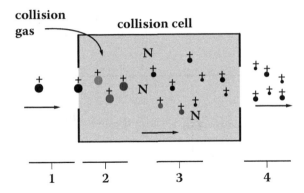

1. **Ions, often a specific *m/z* selected in the first analyzer, are provided with kinetic energy as they are accelerated into the collision cell.**
2. **Ions acquire additional vibrational energy through multiple interactions with collision (reagent) gas atoms.**
3. **Fragmentation occurs when sufficent energy is accumulated by an ion. A second fragmentation may ensue if the initial fragments acquire sufficient additional energy. Neutral species (N) formed are not detected.**
4. **Emergent ions are transfered to the second analyzer.**

FIGURE 3.14   The process of collision-induced dissociation (CID).

The choice of how much kinetic energy to add to the ions depends on both the analyte and the type of data sought. To obtain structural information, a range of energies should be used so that the spectra show progressive, energy-dependent fragmentation; e.g., spectra of the sulfa drug sulfadimethoxine show more extensive fragmentation as the collision energy is increased (Figure 3.15). If the objective is quantification using *selected reaction monitoring* (Section 3.3.3.1), then the collision energy should be adjusted to produce a single, structurally distinctive, fragment ion of high intensity.

The CID process (in positive ion mode) is effective only for protonated molecules. If the *m/z* selected in the first analyzer corresponds to a sodiated species, $[M + Na]^+$, then the stability of these ions results in either no fragmentation, regardless of how much energy is added, or that a point is reached where the acquired energy leads to the complete disintegration of the ion.

When the precursor ions are accelerated with 1–20 kV, the technique is referred to as high-energy CID. The resulting collisions (often with helium rather than argon) excite the electronic state of the analytes and produce a broad distribution of acquired internal (vibrational) energies. The extensive fragmentation provides a great deal of structural information; e.g., high-energy CID can even distinguish leucine from isoleucine

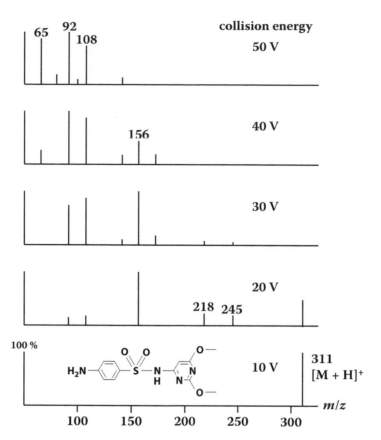

CID spectra of sulfadimethoxine at collision energies of 10–50 volts.
Scanning the voltage that provides the collision energy is used to
   generate multiple spectra with structural information.
A specific voltage, selected to provide a single intense fragment, is
   used in selected reaction monitoring.
The required collision energies are compound-specific and are
   determined empirically.

FIGURE 3.15   Effect of collision energy on collision-induced dissociation (CID) fragmentation patterns.

residues in peptides. High-energy collisions can be carried out only in instruments
that have been modified (or are homemade) for such experiments. Among commercial
instruments, TOF/TOF systems use low kV for CID, but this is dictated, in part, by the
short time that the ions from TOF1 remain in the collision cell.

3.3.2.2   Infrared Multiphoton Dissociation (IRMPD)

Infrared multiphoton dissociation (IRMPD) is used in FT-ICRMS instruments
to generate fragmentations similar to those obtained by CID. IRMPD takes place

within an ICR cell into which an infrared laser is directed through a window. The process is analogous to CID because the analyte ion interacts progressively with multiple photons from the laser until a point is reached where sufficient energy is absorbed to induce fragmentation. The lasers used are carbon dioxide type with a wavelength of 10.6 μm and energies of 0.12 eV/photon.

3.3.2.3   Electron Transfer Dissociation (ETD) and Electron Capture Dissociation (ECD)

There are other means to fragment ions and thus produce MS/MS spectra that contain information different from that obtained by conventional CID. *Electron transfer dissociation* (ETD) and *electron capture dissociation* (ECD) are gaining popularity because they provide structural information that is difficult to obtain, e.g., the preservation of phosphates in the LC-MS/MS analysis of peptides obtained from the enzymatic digestion of phosphoproteins.

ETD is used with QqQ, QTOF, and orbitrap instruments. Reagent species, such as fluoranthene and nitrosobenzene, are activated by conversion to their respective radical anions in a corona discharge. The radicals are then mixed with the analyte (that has been ionized using ESI), resulting in the transfer of an electron from the reagent to the analyte. The acquired energy promotes fragmentation pathways that are different, shown in Section 3.5.1.5, from those seen in CID.

ECD, used with FT-ICRMS instruments, is an alternative way to produce fragmentations similar to those obtained by ETD. The objective is still to add an electron to the analyte, but here the necessary interactions occur within the ICR cell and utilize thermal electrons (<0.2 eV) emanating from a filament.

3.3.2.4   Photodissociation and Surface-Induced Dissociation

Ions containing a chromophore that absorbs energy at a specific wavelength may undergo electronic excitation when irradiated with photons of well-defined energy, provided by a UV laser. *Photodissociation* (PD) is highly selective because only those ions, or specific parts of ions, are activated that can absorb energy at the wavelength of the irradiating laser. Trapped-ion instruments provide a desirable environment for PD because ions are held for prolonged periods in regions where they can be irradiated. The IRMPD method (Section 3.3.2.2) is also based on a PD process, but it is not structure dependent and is more generally applicable because of the wavelength utilized and the energies of the photons.

In *surface-induced dissociation* (SID) ions gain internal energy and subsequently decompose upon collision with a *target surface* that is positioned perpendicular to, or at an angle to, the ion path. In an alternative approach, ions pass through very narrow channels where bouncing off the inner surfaces provides the additional energy necessary to fragment the ions. The composition and physical nature of the surface are major variables that affect the extent of fragmentation. SID is efficient because even low-energy collisions provide analytes with high internal energies with narrow energy distributions. An application of SID is to obtain information on energetic aspects of the fragmentation mechanism of protonated peptides.

3.3.3 TANDEM-IN-SPACE MS/MS

In *tandem-in-space* MS/MS a collision cell is placed between two analyzers that may be the same, e.g., QqQ or TOF/TOF, or a combination of different types, e.g., QTOF or LT-orbitrap. The analyzers are separated *in space* in all such instruments.

3.3.3.1 MS/MS Techniques Using Triple Quadrupole (QqQ) Analyzers

The triple quadrupole, QqQ, is the only member of the first generation of multiana-lyzer instruments still available commercially (other early MS/MS instruments were constructed using magnetic sector technology). The QqQ is still the most frequently purchased multianalyzer system. In terms of analytical performance, the name triple quadrupole is a misnomer because only the outer two quadrupoles (Q1 and Q3) can be scanned. The middle quadrupole (q2) has no analytical function but serves as the collision cell where ions emerging from Q1 undergo CID. An rf-only field is used in q2 to constrain the fragment ions derived from the collisions, and to transmit them into Q3 without any separation. In fact, q2 (the collision cell) does not have to be a quadrupole but can be a chamber in which collisions are used to induce fragmenta-tion of the analyte ions. The cell should be designed to prevent the scattering of the fragment ions as they are formed and transported to Q3.

Quadrupoles are scanning analyzers with limited sensitivity for collecting full spectra, but they are highly sensitive when monitoring a single *m/z* value (Section 2.3.1). Using two analyzers that may be either scanned or held static provides the versatility of triple quadrupoles. The possible combinations of scanning and static operation permit several modes of data collection that enable significantly different analytical strategies, as described below and shown in Figure 3.16.

3.3.3.1.1 *Product Ion Scanning Mode*

Q1 static, Q3 scanned. Ions (*m/z*) of interest are selected in Q1 followed by transfer into the collision cell (q2), where they undergo CID through interaction with the

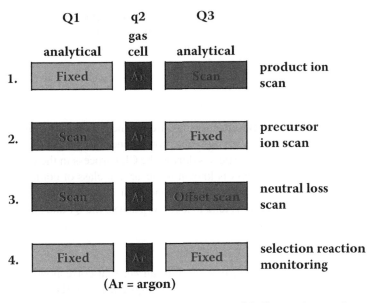

Q1	q2 gas cell	Q3	
analytical	cell	analytical	
1. Fixed	Ar	Scan	product ion scan
2. Scan	Ar	Fixed	precursor ion scan
3. Scan	Ar	Offset scan	neutral loss scan
4. Fixed	Ar	Fixed	selection reaction monitoring

(Ar = argon)

1. **Product ion scans (in Q3) provide structural information on ions with a specified m/z, selected in Q1.**
2. **Precursor ion scans (in Q1) determine the precursor ions that yield the product ion (m/z) selected in Q3.**
3. **Neutral loss scans are used to find 'families' of related compounds by scanning both Q1 and Q3. The analyzers are offset (from each other) by an m/z corresponding to a common neutral structural subgroup that is lost during CID.**
4. **Selected reaction monitoring (SRM) is used for quantification. Improved signal:noise ratios at the selected m/z are observed in Q3 because the CID process suppresses the chemical noise.**

FIGURE 3.16   Modes of operation of triple quadrupole instruments.

neutral atoms (molecules) of a target (reagent) gas, usually argon. The *product ions* formed in q2 are analyzed according to their m/z in Q3, producing a mass spectrum of the ions derived from the selected precursor ion. This is the most common MS/MS strategy for obtaining structural information about unknown compounds. The data are similar to those generated using other multianalyzer instruments (see later); however, sensitivity is one to two orders of magnitude lower than that available with nonscanning analyzers (because Q3 is a scanning analyzer).

3.3.3.1.2   Precursor Ion Scanning Mode
Q1 scanned, Q3 static. This approach is the reverse of product ion scanning. Q1 is scanned first, and the resulting ions pass into q2, where they undergo CID. Q3 is then held static, allowing passage of only a single preselected m/z value. The signal in Q3 can be related to the scan function of Q1; thus, a spectrum is acquired of the

precursor ions that have dissociated into the specified product. An important application of this strategy is the identification of the components of mixtures that belong to a specific class of compound. Precursor scans can be used to search for a desired class of compounds in complex mixtures because Q3 is set to select a characteristic, charged CID fragment.

3.3.3.1.3 *Constant Neutral Loss Scanning Mode*

Q1 scanned, Q3 scanned. Both Q1 and Q3 are scanned at the same rate, but the scans are offset from each other by an *m/z* value corresponding to a specific, selected, component that is lost as a neutral species during the CID process in the collision cell (q2). When a particular *neutral loss* is known to occur in a class of compounds, this type of MS/MS analysis will identify the components of mixtures that release the neutral fragment; e.g., carboxylic acids tend to lose CO_2, and so the Q1 and Q3 scans are offset by 44 Da.

3.3.3.1.4 *Selected Reaction Monitoring (SRM) Mode*

Q1 static, Q3 static. Q1 is set to pass only an *m/z* that corresponds to the analyte, usually the $[M + H]^+$ ion. Ions emerging from Q1 undergo CID in q2, and a specific product ion (*m/z*) is monitored using Q3.

Ions emerging from Q1 include not only those from the compound of interest, but also ions from all other species present that produce the same *m/z*. Such ions are the *chemical background/noise* that can have multiple origins, e.g., co-extracted compounds from the samples, ions derived from the extraction process itself (such as solvent contaminants), or compounds that are resident within the instrument. As samples with progressively lower analyte concentrations are analyzed, the chemical noise occurring at the *m/z* of interest will eventually limit the ability to differentiate the analyte ions from background noise, i.e., the *detection limit* is reached. If a detector is placed after Q1, the limitation resulting from the chemical noise corresponds to that of single quadrupole instruments (Figure 3.17, left side). In contrast, in triple quadrupoles the CID in q2 creates fragments, and it is unlikely that both the analyte and interfering noise ions will generate the same product ion *m/z*. Therefore, when Q3 is set to monitor a specific product ion of the analyte, the background ions (noise) from interfering species, which have been fragmented to other *m/z* values, will be rejected. Although analyte ions are lost in this process, there still is a substantial benefit because a greater suppression of the background ions takes place, and thus the signal-to-noise ratio is improved (Figure 3.17, right side). The specificity of the data is reduced because full spectra are not collected. However, specificity can be increased by monitoring a second fragment ion from the analyte and ensuring that the ratio of the intensities of the two ions in the sample extract is the same as that observed for a standard of the analyte.

SRM is the MS/MS acquisition mode analogous to selected ion monitoring (SIM) in single-analyzer instruments. When both precursor and product masses are known to be characteristic of the analyte SRM permits quantification of target (known) compounds with high sensitivity and concurrent high specificity, even in highly complex biological or environmental matrices. Applications include obtaining pharmacokinetic information on drugs and their metabolites, and quantification of endogenous

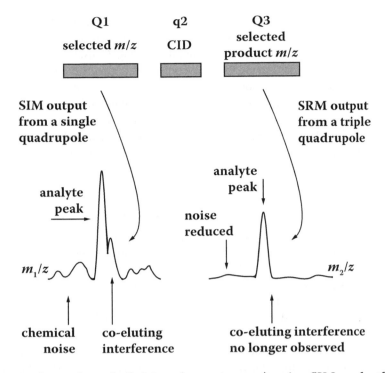

In a single quadrupole (left ion chromatogram), using SIM mode, the chemical noise and/or presence of coeluting compounds limits the attainable signal to noise ratio (S/N) and therefore the limit of detection.

In a triple quadrupole (right ion chromatogram), using SRM mode, the signal-to-noise ratio is improved, even though ions are lost during the collision-induced dissociation (CID) in q2 . The improvement is due to the low probability that interfering species (chemical noise and coeluting compounds seen in the output from a single quadrupole) will have the same CID progeny ion as that selected for the analyte, and so they will not be detected in Q3.

FIGURE 3.17 Comparison of ion traces from single and triple quadrupole analyzers.

compounds as well as environmental contaminants. Because in such applications analytes are identified and quantified in highly complex matrices such as blood plasma, urine, organ tissue, environmental sediments, or plant tissues, some preliminary separation, usually by chromatography, is needed. The chromatography is fast when ultra-performance liquid chromatography (UPLC) is available using short (5 cm) columns and a 1 or 2 min solvent ramp that provides a basic separation, particularly the removal of inorganic salts. Optimal separation is not obtained with a fast ramp; however, the specificity provided by the combination of CID and Q3 selectivity more than compensates for this problem. The strategy of combining fast chromatography with SRM is a major advantage in applications where a large number of samples need to be analyzed rapidly and where the rate-limiting step is the chromatography, e.g., in drug development, biomarker research, and environmental monitoring. Consider the thousands of analyses that must be made during the development of a drug. A thousand samples at 30 min per run would require some 21 days of continuous instrument operation but only about 2 days if each chromatographic run is reduced to 3 min.

SRM is one of the most common uses of QqQ instruments because both analyzers are utilized to their maximum sensitivity, thereby exceeding the limits of other instrument formats because of the duty cycle benefits (Section 2.3.1).

3.3.3.2   MS/MS Techniques Using Time-of-Flight/ Time-of-Flight (TOF/TOF) Analyzers

In TOF/TOF instruments the ions produced by MALDI are separated in TOF1, and an ion of interest is selected based on its expected time of exit from TOF1 using a *timed ion selector* (TIS) that uses high-speed electronics to deflect ions arriving before and after the analyte ion. The rate at which the electronic gate operates and the speed at which ions travel result in ions arriving at the collision cell in a window (time period) that is several mass units wide. The selected ions are fragmented, followed by analysis of product ion spectra in the second, reflectron-equipped TOF. Because the ion selection window (in the TIS) is several mass units wide, the isotopes of the analyte are also included in the CID process; thus, the fragments will have their respective isotope patterns to assist in their identification. Unlike quadrupole-based MS/MS instruments, where only nominal mass data are acquired, TOF/TOF systems provide accurate masses for the product ions.

TOF/TOF is the most restrictive of the MS/MS formats because neither analyzer can be scanned. However, these instruments combine the high sensitivity associated with MALDI, with CID conducted at keV energies. In TOF/TOF instruments the residence time of the ions in the CID cell is shorter and lower collision gas pressures are used than in other MS/MS systems. CID takes place at 1–2 keV to provide sufficient fragmentation, and this has the added benefit of providing fragmentation that is less structure dependent and more comprehensive than that obtained with the 10–100 eV CID used in most MS/MS systems.

3.3.3.3   MS/MS Techniques Using Hybrid Instruments

Hybrid instruments consist of a quadrupole or quadrupole-based trap followed by a TOF or an FT-MS type trap as the second analyzer. The spectra generated are usually from product ion scanning. How these spectra are acquired usually depends on the speed of operation of the second analyzer.

TOF analyzers are the fastest instruments for the collection of high-resolution spectra. The current generation of TOF analyzers is capable of collecting multiple spectra on a chromatographic timescale, where peaks are only a few seconds wide, at resolutions of 20,000–40,000 (or even 60,000) with accurate mass measurement. On the other hand, there is a progressive reduction of the resolution attainable on FT ion traps as the speed of the production of spectra is increased (Section 2.3.4.2). This has led to an alternative MS/MS strategy: a portion of the molecular ion signal is passed to the FT trap (typically an orbitrap) to obtain data for accurate mass determination. The remaining molecular ions are retained in the initial analyzer (usually a linear ion trap with its own detector), where they are fragmented and then scanned to provide spectra of the CID products. Such spectra have unit resolution and do not provide accurate mass data.

When an FT-IRCMS instrument is not time limited, i.e., when direct sample introduction rather than chromatography is used, spectra are collected in the ICR cell where the MS/MS data are produced using a series of rf pulses and fragmentation processes. The pulses enable the selection of the m/z of interest while ejecting all other ions from the cell. The kinetic energy of the selected precursor is increased (also with an rf pulse) and then trapped for several milliseconds, during which MS/MS fragments are obtained through collisions with a target gas (CID) or by using IRMPD (or other, more specific photodissociation methods). The fragmentation takes place under low-energy conditions that involve multiple collisions, or interactions with multiple photons. The product ions are then excited into coherent motion with a further rf pulse, and finally, the ion image currents are recorded for subsequent FT analysis. Advantages include the opportunity to control the energy spread of the ions through cooling and reacceleration, the ability to analyze product ions at very high resolution, and the option to carry out various protocols for selection, isolation, fragmentation, and analysis, multiple times, depending on the ion current available.

3.3.3.4 Data Dependent Acquisition (DDA) and CID Fragmentation by Voltage Switching (Data Independent Acquisition, DIA)

When multiple unknown species are analyzed by LC-MS, e.g., a tryptic digest of a protein, numerous protonated molecules are entering Q and real-time decisions are needed as to which ions to select for MS/MS. The first step of the strategy is to send all ions emerging from the source directly to the TOF analyzer to obtain molecular mass information on all compounds present. Next, software is used to initiate and control a series of subsequent steps: (1) select a specific ion, (2) set the Q voltages to allow passage of that ion, (3) control the CID process, and (4) collect data for the resulting fragments separated in the TOF analyzer. This ion selection cycle, also called *data-dependent analysis*, is a continuous process. It is often possible to monitor the fragmentation of a number of m/z values during the elution of a peak from the LC. A difficulty with DDA is that the ions selected for fragmentation are usually the most intense ones in the spectrum that is likely to contain several protonated species. Therefore, other co-eluting species of lesser intensity will not be analyzed by MS/MS, even though the minor components of the sample are often of biological importance (Figure 3.18).

Because LC chromatograms are often highly complex, with multiple peaks co-eluting, there is a need to increase the number of ions for which MS/MS data are obtained. DIA provides improved coverage by using the quadrupole in the wide band pass mode, allowing all ions to enter the collision cell, and cycling the voltage that propels the ions into the cell; thus enabling the collection of both molecular mass and fragmentation information on all protonated species. When low voltages (e.g., 5 V)

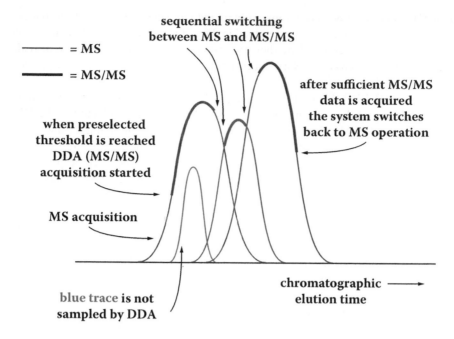

FIGURE 3.18   Data-dependent acquisition (DDA) operation.

are used, the ions acquire insufficient energy to undergo CID, i.e., remain intact, and molecular mass data are recorded. Rapidly raising the voltage (e.g., using a 20 to 50 V scan over 1 s) increases the energy given to the ions, initiating CID (fragments are formed) and enabling the acquisition of structural information. Molecular mass and fragmentation data are generated, essentially simultaneously, because of the fast (every second) switching between the low and high voltages (faster than the time scale of LC peak elution). If more than one compound is present, mixed spectra are produced and sophisticated software is required to determine which fragments are derived from which protonated molecule. Even a slight difference in the elution profiles of two ions indicates that they originate from different compounds. Accurate mass information is essential to rationalize whether two product ions have different precursors.

3.3.4 MS/MS Techniques Using Tandem-in-Time (Trapped-Ion) Analyzers

In *tandem-in-time* mass spectrometers all operations take place within a single-ion storage device, but at different times. The most common tandem-in-time instrument is the *quadrupole ion trap* (QIT). The ongoing development of a higher-sensitivity counterpart, the *linear ion trap* (LIT), will presumably lead to an eclipsing of the QIT.

There are four basic steps in tandem-in-time MS/MS analyses: (1) selection of the mass of a precursor ion from the analyte, (2) application of an excitation voltage to the end cap electrodes, and injection of the collision (reagent) gas, (3) dissociation of the energized ions to form product ions, and (4) mass analysis of the product ions. During steps 2 and 3 the selected ions are accelerated momentarily to higher kinetic energies by the excitation waveform in the presence of the collision gas. The process, which lasts for ~1 ms, leads to multiple collisions and other low-energy processes that produce spectra containing a limited number of fragment ion types. Lasers may be used instead of the collision gas to produce photodissociation products.

The selection, fragmentation, and scan process is rapid and completely under electronic control. The process can be undertaken multiple times to acquire MS^n spectra because everything occurs within a single analyzer. Obtaining MS^n data is a distinct and important capability of QIT and LIT analyzers (Figure 3.19). In practice, n is usually 3 or 4 because of ion losses that occur during each cycle. This limitation is particularly relevant to the QIT, where the number of ions that can be allowed initially into the trap must be limited to around 2,000 to avoid space:charge problems when collecting full spectra. In contrast, spectra can be obtained using an LIT even when there are about 30,000 ions present. Availability of MS^n is an asset that allows the detailed exploration of a compound's structure because additional information is provided by the spectra obtained at each repetition (Figure 3.19).

3.3.5 Quantification of Small Molecules

Despite the fact that mass spectrometers are the predominant instruments in many laboratories where quantification is the primary objective, a common comment concerning mass spectrometry is that it is not a good quantification technique. There is also the suggestion that the gold standard for quantification is LC combined with UV detection. This conjecture is questionable, as the chromophores of compounds are highly variable; e.g., the UV spectra of alkanes and aromatic compounds are very different, with the former detectable only at short wavelengths (low 200 nm), whereas the latter are observed in the mid-200 nm region. An additional complication with UV detection is the increasing background interference that occurs as the selected wavelength approaches 200 nm because of absorption by the elution solvents. Mass spectrometry is also preferable to UV detection in terms of selectivity; assaying for a characteristic mass is much more diagnostic than monitoring a UV wavelength.

It can be argued that differential responses are problematic in mass spectrometry and that they depend on the structure of the analyte, especially when ESI is used. ESI reflects the polarity of the analyte, meaning that instrument response will favor polar species. APCI is more uniform in this respect but has its own problems, including excessive fragmentation.

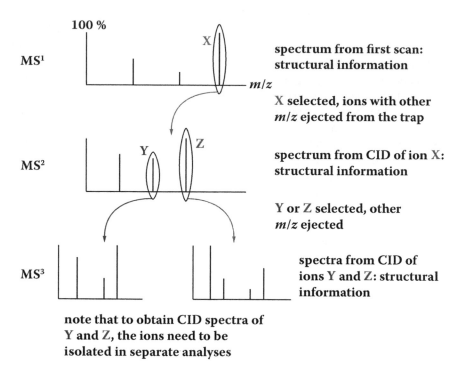

100 %

MS¹

X

spectrum from first scan:
structural information

m/z

X selected, ions with other
m/z ejected from the trap

MS²

Y   Z

spectrum from CID of ion X:
structural information

Y or Z selected, other
m/z ejected

MS³

spectra from CID of
ions Y and Z: structural
information

note that to obtain CID spectra of
Y and Z, the ions need to be
isolated in separate analyses

In a QIT (LIT) the same analyzer is used sequentially, in time, to
generate spectra.
An ion isolated from the MS1 scan undergoes collision-induced
dissociation (CID) yielding a product ion spectrum that
provides structural information on the selected ion, MS2.
The process of isolation:CID:scan can be repeated to give
additional data, MSⁿ spectra.

FIGURE 3.19   MSⁿ data generation using a quadrupole type ion trap (QIT/LIT).

Overall, in quantification, it is not the method of detection that is important
(e.g., GC-FID, LC-UV, GC-MS, LC-MS), but how the data are standardized with
appropriate calibration curves. The most important MS system for the quantification
of small molecules is the triple quadrupole mass spectrometer, particularly when
used in SRM mode (Section 3.3.3.1). Although much of what is described below
concerns small molecules (for techniques used in the quantification of proteins see
Section 3.5.1.9), the same basic considerations apply for all targeted analytes.

A *calibration curve*, also called a *standard line*, is a plot of a series of numbers
that represents the observed signals from a set of samples with known concentrations
of the analyte (y-axis) plotted against those concentrations (x-axis). There are three
common methods of quantification: external standard, internal standard, and standard addition.

The *external standard* method uses a calibration curve based on a series of individual samples of a matrix spiked with increasingly higher concentrations of the analyte as well as a *blank* sample to which no analyte has been added. A *matrix* is the material in which the analyte is present, e.g., plasma, urine, water, sediment, or even air. Blanks must be used in both qualitative and quantitative analyses to make certain that no reagent in the analytical protocol contributes (or interferes with) the measurement. It is important to realize that any blank (especially biological matrices) may contain *endogenous* concentrations of an analyte. The same methodology should be used for the extraction and analysis of the calibration standards and the samples containing the analyte at unknown concentrations. A calibration curve (which is typically linear) is obtained when the resulting signal intensities for the calibration samples are plotted against the known analyte concentrations (Figure 3.20). The slope of the line is described by a linear regression equation: $y = mx + c$, where y = measured signal intensity at a specific m/z value, m = the slope of the line, x = known concentration of the analyte, and c = a constant given by the intercept of the line with the y-axis. To derive unknown concentrations of the analyte (x-values), the intensities of the signals obtained for the analyte are inserted into the equation as the y-values.

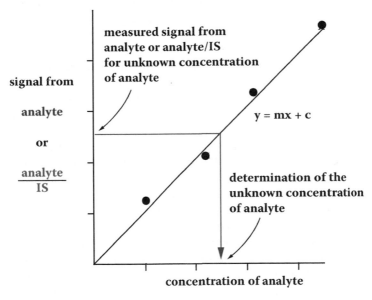

Calibration curves are similar whether an external or internal standard methodology is used.

The difference is that in the external standard method the y-axis corresponds to the signal strengths of the analyte while in the internal standard method the y-axis is comprised of the ratios of the signals from the analyte and internal standard (IS).

The unknown concentration of an analyte is obtained from the linear regression equation of the calibration curve.

FIGURE 3.20   Calibration curves for external or internal standard methods.

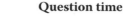

Question time

What is a matrix?

It is any medium from which an analyte is
   extracted prior to analysis, e.g., plasma, tissue,
   urine, sediment, air, water, and even a coelacanth.

The matrix in MALDI has a different meaning.
   It refers to a medium from which a
   sample is ionized.

While convenient, especially if sample preparation is simple and an autosampler is used so that injections are reproducible, the external standard method does not account for sample variability. Take, for instance, two samples, A and B, in both of which an analyte is present at the same concentration. If the efficiency of extraction of the analyte is 100% from sample A but only 70% from sample B, then the signal intensities will vary by the same ratio (Figure 3.22a). When these values are inserted into the equation of the calibration curve, the calculated concentrations will vary by the same 30% that the extraction efficiencies differ, even though the concentrations of the analyte are the same in both samples.

The *internal standard* (IS) method compensates for the recovery problem encountered with external standardization by adding a compound (the IS) of known concentration to each sample prior to extraction. The compound chosen as the IS should have the same extraction efficiency, chromatographic retention time, and mass spectrometric properties as the analyte. These properties can be provided by an IS that has a chemical structure similar to that of the analyte.

IS-based calibration curves are prepared the same way as those based on external standardization, by using a series of samples with increasing concentrations of the analyte. However, the methodology uses a known, constant amount of the IS, which is added to each sample prior to extraction. The signals for IS and the analyte are obtained (Figure 3.21), but here the ratios of the peak areas of the analyte to the IS are plotted against the known concentrations of the analyte (Figure 3.20).

The benefit of adding the same, constant amount of IS to all samples, with either known or unknown concentrations of the analyte, is seen in the case of the two samples discussed above, A and B, both of which have the same concentration of analyte but different recoveries. The inclusion of an IS prior to the extraction step results in better accuracy (Figure 3.22b). The basis of the improvement is that the IS is similar to the analyte (essentially identical when the IS is isotopically labeled), so that its behavior is the same (or very similar) to that of the analyte during the extraction, chromatographic, and mass spectrometric procedures. Therefore, in sample B,

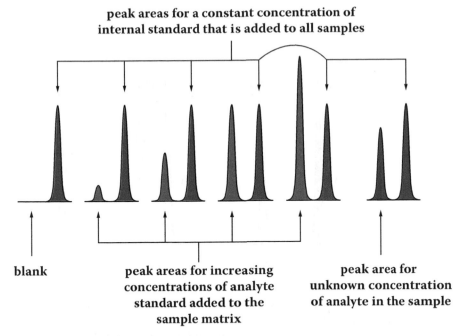

**peak areas for a constant concentration of
internal standard that is added to all samples**

blank

**peak areas for increasing
concentrations of analyte
standard added to the
sample matrix**

**peak area for
unknown concentration
of analyte in the sample**

Ratios of peak areas for the analyte : internal standard (both are
added to the sample matrix and extracted) are plotted against
the concentrations of the analyte, and a regression equation
is calculated.

The same analyte : internal standard peak ratios are obtained for
samples with unknown analyte concentrations. Concentrations
are determined by comparison with the calibration curve using
the regression equation.

FIGURE 3.21 Internal standard calibration method data.

the 70% recovery will apply to both analyte and IS, and thus the ratio of the signals
will be the same as that for sample A, where the recovery was 100% (Figure 3.22b).
The calculated concentrations will be the same for both samples when the ratio is
inserted (as the y-value) into the equation for the calibration curve. This shows how
incorporating an IS provides an improvement over external standardization.

The structural similarity of the IS and the analyte may result in their co-elution
from the chromatograph, particularly when the IS is (stable) isotopically labeled. Such
co-elution is unsatisfactory when the detection is, for instance, by UV (in LC) or by
flame ionization (in GC) because the analyte and IS must be separated in time. In
mass spectrometry the IS can be separated from the analyte in a second dimension,
with mass difference being used for characterization rather than just chromatographic
retention time (Figure 3.23). Separation based on mass is the basis of using isotopically
labeled analogs of the analyte as the IS. The physical/chemical properties of such a

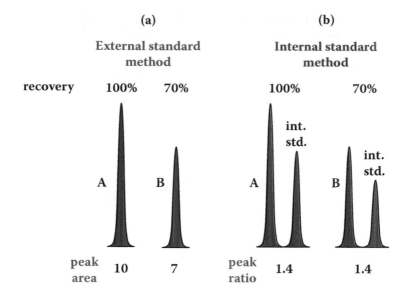

	(a)	(b)
	External standard method	**Internal standard method**

Samples A and B are present at the same concentration but the efficiencies of their extractions differ.

(a) The external standard method does not account for differences in recovery. The detected peak areas for A and B differ and, consequently, so does the apparent analyte concentration.

(b) The internal standard method uses the ratio of the peak areas for the analyte and internal standard. Because the extraction efficiencies are the same for both the analyte and the internal standard, the area ratios obtained for samples A and B are the same. This accounts for the differences in recovery and provides an accurate determination of the analyte concentration.

FIGURE 3.22 Comparison of the effects of analyte recovery on the accuracy of the methodology.

standard are essentially identical to those of the analyte. For small molecules (such as drugs) the molecular mass of the isotopically labeled IS should be at least three mass units higher to avoid interference from the ^{13}C isotope contribution of the analyte. For instance, isotopic labeling of a small molecule drug could replace a $-CH_3$ group with a $-CD_3$ group. The same result would be achieved by replacing three ^{12}C with three ^{13}C, but at considerably higher expense. For analytes of higher molecular mass, e.g., peptides, the larger number of carbon atoms means increases in the number of ^{13}C atoms and expansion of the isotope envelope and, consequently, the IS must have increased numbers of deuterons (or ^{13}C) to avoid interferences.

Calibration curves do not always pass through zero (the origin), nor are they always linear (Figure 3.24). Possible reasons for a line not passing zero (Figure 3.24a) are that the analyte is already present in the sample, i.e., it is *endogenous*, or that there

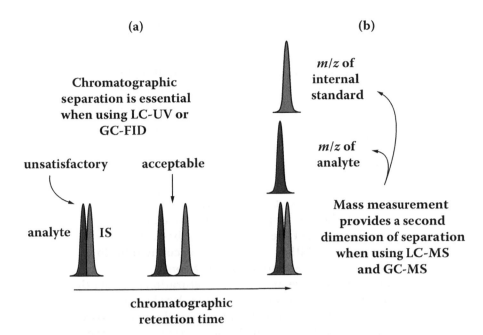

(a)

(b)

Chromatographic
separation is essential
when using LC-UV or
GC-FID

m/z of
internal
standard

m/z of
analyte

unsatisfactory   acceptable

analyte   IS

Mass measurement
provides a second
dimension of separation
when using LC-MS
and GC-MS

chromatographic
retention time

(a) Without mass spectrometry, e.g., when using LC-UV or GC-FID,
     peaks must be separated chromatographically to obtain peak
     areas for the analyte and internal standard.
(b) When LC or GC is coupled with MS, different *m/z* values can
     be used to deconvolute overlapping chromatographic peaks
     and allow the use of isotopically labeled internal standards.

FIGURE 3.23 Chromatographic vs. mass separation.

is a second compound at the same chromatographic retention time contributing to the
signal. In the former case the calculated concentration observed is correct, i.e., includes
the endogenous concentration of the analyte. In the second case the concentration has
to be adjusted to compensate for the interference. The likelihood of encountering interfering compounds is reduced with triple quadrupoles (Figure 3.17).

A nonlinear response occurring as the concentration of the analyte is increased is
illustrated in Figure 3.24b. This may be due to variations in ionization efficiency at
higher analyte concentrations or to a saturation of the detector response. When quantifying peptides, concentration changes can cause alterations in the charge states
of the peptide ions and, as a result, nonlinearity if a single ion is being monitored.
There are two approaches to dealing with nonlinear curves: either a quadratic equation is fitted to the data to accommodate the fall-off in response (often required for
peptides), or sample sizes need to be adjusted so that only the linear portion of the
response is used.

Using the *standard addition* method can further increase the accuracy of the
determination of concentrations. Sufficient sample quantity must be available

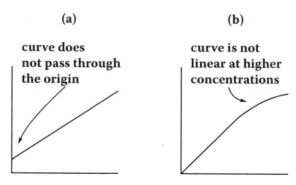

(a)   (b)

curve does   **curve is not**
not pass through   **linear at higher**
the origin   **concentrations**

(a) **A calibration curve that does not pass through the**
     origin indicates the presence of (i) an endogenous
     concentration of the analyte, or (ii) background
     interference. Either case must be accounted for in
     determining the analyte concentation.
(b) **A nonlinear curve may indicate saturation of either the**
     ionization process or the detector. A polynomial
     equation may be fitted to the data or the sample size
     adjusted so that only the linear portion of the curve
     is used.

FIGURE 3.24   Calibration curves needing extra consideration.

because multiple aliquots are required each time a concentration is determined.
Sample preparation for the standard addition method uses a set of aliquots of the
sample in the following manner: the first aliquot is used as is, but each subsequent
aliquot is supplemented with an increasing amount of the analyte. The analyte is then
extracted from each sample, analyzed, and a graph is plotted of signal vs. amount of
analyte added (Figure 3.25). In essence, a calibration curve is made for every sample.
The intercept of the curve with the negative side of the x-axis (when expressed as a
positive number) gives the concentration of the analyte. An advantage of the standard addition method is that no IS is required because the analyte is used as its own
standard. Although this strategy is very accurate, it requires a sufficient quantity of
the specimen to prepare multiple samples. Furthermore, additional instrument time
is required for the analysis of the multiple aliquots of each sample.

Of the three quantification methods, the internal standard method is often favored
because it provides both accuracy and speed. Although the use of an IS may compensate for variations in recovery, it does not provide a measure of absolute recovery.
The latter can be estimated by comparing the signal from an analyte in the extracted
sample with that from a known amount of the same compound injected independently, or by adding a second standard (in addition to the IS) to the sample immediately prior to injection onto the chromatographic column. The latter is the approach
adopted by the U.S. Environmental Protection Agency.

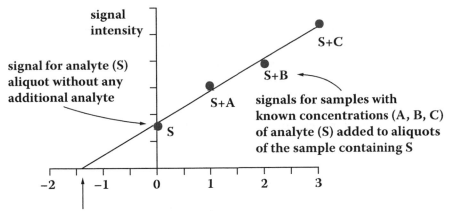

intersection with the x-axis gives the
concentration of the unknown, S, when
stated as a positive number,
i.e., by disregarding the negative sign of
the axis (a negative concentration is
not possible)

**Standard addition uses multiple sample aliquots to each of which are
added progressively increasing amounts of the (pure) analyte.**

FIGURE 3.25 Analyte concentration determination using the standard addition method.

3.3.5.1 Standard Deviations and Method Validation

When at least three samples (preferably more) with known concentrations (standards)
are analyzed at each concentration, standard deviations can be calculated from the
peak areas (external standard and standard addition methods) or peak area ratios
(internal standard method), thereby determining the confidence of data for unknown
concentrations. The standard deviations are usually plotted as error bars at each point
on the calibration graphs.

Associated with calibration is the assessment of the validity of the methodology, known as *analytical method validation* (AMV). Categories include *repeatabil-
ity, intermediate precision,* and *reproducibility*. The first two provide a measure of
whether results are consistent intra- and interday from the same laboratory, while
reproducibility reflects the ability to transfer an assay to another laboratory. Other
AMV categories include the accuracy, linearity, and concentration range over which
an assay is effective. The term *robustness* is used as a measure of how slight variations in the method affect results, e.g., a change in a chromatographic gradient.

3.4 OMICS

Omics is a *neologism*, a newly coined term that is entering common use but is not yet
accepted fully in mainstream language. In molecular biology, omics refers (rather

informally) to specific fields of study, such as proteomics, genomics, metabolomics, lipidomics, and glycomics. As these names imply, omics covers both large and small molecules. The term is not limited to biology; e.g., petroleomics refers to the study of crude oil. Omics is now used to characterize, sometimes in a strange (even comic) manner, the study of almost any field, e.g., cellomics, hygienomics, and enviromics. These names not only satisfy scientists' love of categorizing information, but more importantly, emphasize the simultaneous study of all components and interactions (biochemical, structural, etc.) of entire systems of interest.

The suffix *-ome* refers to the *totality* of something with respect to its study within an omics. For example, the term *proteome* (a noun) refers to the proteins of a cell, tissue, or organism considered collectively in cellular and molecular biology. Note that in medicine, *-ome* (or *-oma*), as part of a noun, refers to tumors or swellings.

Mass spectrometry has recently acquired a major role in omics applications, particularly in proteomics and metabolomics, along with lipidomics, which may be considered a subset of the latter, and glycomics, which bridges both. Further sub-categorization in proteomics is based on characteristics of the proteins, such as the types of post-translational modifications, e.g., phosphoproteomics. One reason for the rise of the prominence of omics is that there have been major advances in high-throughput analyses (at reasonable cost), facilitating the rapid collection of spectra (and often concentrations) for very large numbers of analytes with similar compositions or structures. Compiled data collections have been complemented with computerized informatics techniques that make it possible not only to identify and quantify the analytes, but also to establish, and statistically confirm, structural and functional relationships.

Despite the truly mind-boggling instrumentation and techniques of analytomics (a comic omic?), mass spectrometers are still being used primarily to determine both the *m/z* values of ions, for identification or confirmation (qualitative analysis), and their abundances (quantitative analysis).

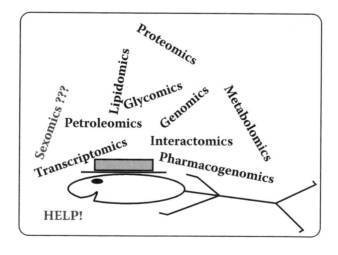

3.4.1 METABOLOMICS

Mass spectrometry has long been associated with the identification, concentration, and fate of compounds of relatively low molecular mass in biological systems, including endogenous constituents as well as pharmaceuticals and other metabolites. Such studies have continued to expand with the development of ESI and APCI sources, that allow the analysis of polar molecules without derivatization.

The aim of *metabolomics* is to determine the interactions between *all* small molecules present in entire biological systems, with the ultimate objectives of describing and understanding how these interactions lead to the functioning and behavior of the systems involved, i.e., a *systems biology* approach. Metabolomics also includes the investigation of the relationships or interactions between individual organisms and their environments, e.g., nutrition, toxicity, and pathology. (A related, and somewhat confusing, term, *metabonomics*, refers to the study of the response of living systems to pathophysiological stimuli or genetic modification.)

Metabolomics includes both the qualitative and quantitative changes (*metabolism*) of small molecules of endogenous and exogenous origin within a living biological system, such as a cell, tissue, or organism. There are two categories of metabolism, *catabolism*, which deals with the breakdown of organic matter, e.g., to obtain energy from nutrition or to remove foreign (potentially toxic) substances, and *anabolism*, during which cell components are constructed, including biopolymers. Both the products and intermediates of metabolism are called *metabolites*, and the series of transformative steps (usually mediated by enzymes) for a given set of compounds is a *metabolic pathway*. The *metabolome* is the entire set of small molecule metabolites present—at a given moment—in a cell, tissue, organ, biofluid, or entire organism. Understanding and monitoring the changes in the metabolic profile may provide the opportunity to better understand normal or disease states as well as the consequences of compounds encountered through nutrition and environmental exposure.

Mass spectrometry, often combined with GC, LC, or CE, is the key technique for the identification and quantification of metabolites. Recovery of compound classes for analysis can involve liquid and solid phase extraction or the direct introduction of cell components. Advanced, sophisticated sample introduction techniques include the profiling of laser-dissected tissues and sampling of single cells using microcapillaries. All sample introduction approaches enable the characterization and quantification, including the study of differential accumulation, of compounds in specific tissues or cells. *Imaging mass spectrometry* (Section 3.6) is used for the spatial visualization and quantification of individual metabolites across selected tissues, leading to an understanding of the functional roles of individual metabolites. Other important analytical methods utilized in metabolomics include nuclear magnetic resonance (NMR) spectroscopy and laser-induced fluorescence (LIF).

The versatility of mass spectrometry for metabolomic studies is based on the different sample introduction techniques, the various ionization methods, the ability to collect both MS and MS/MS spectra frequently, including accurate mass measurement, as well as performing SIM and SRM (Section 3.3.3.1). Acquisition of MS/MS spectra and their incorporation in the construction of searchable libraries, which include different fragmentation and structural information from the libraries of EI spectra, has acquired major importance because the soft ionization methods give little structural information without MS/MS. A large proportion of the metabolites detected in large-scale studies remain unidentified even though significant progress has been made in establishing MS/MS libraries, using authentic standards, that can be used to match unknown analytes. Characterization of unknown compounds remains a challenging problem that is now being addressed, in part, by computational (in silico) mass spectrometry.

In nontargeted, large-scale profiling studies, relative quantification is used, based on normalization with respect to internal standards or selected (known) metabolites. In targeted metabolomic studies, absolute quantification is the objective and is accomplished by using internal standards, preferably isotopically labeled forms, which are matched to known metabolites. An alternative strategy uses stable isotope *tags* introduced metabolically through the use of a labeled feedstock to circumvent the difficulties of synthesizing large numbers of labeled metabolites. Such tags can be tracked using a combination of chromatographic retention times and the known offset between compounds (with known natural isotopic abundance) and a labeled analog. A major limitation is that the currently available dynamic range (recovery/instrumental) of ~10^6 is a long way from the estimated 10^{12} required to cover the expected concentrations of potentially important metabolites in a given biological system.

3.4.2 LIPIDOMICS

The range of lipids that can be identified and quantified today is a far cry from the fatty acid methyl esters injected onto GC-MS systems 30 years ago. *Lipidomics* aims to understand the pathways and networks of simple and complex lipids, as well as families of lipids, in cells, tissues, and entire organisms, seeking to characterize, qualitatively and quantitatively, their biological significances, in both health and disease. Lipidomics is a subset of metabolomics given that the latter concerns the interrelationships of all small molecules in a biological system. The *lipidome*, the complete lipid profile of a eukaryotic cell, tissue, or organ, may contain tens of thousands of compounds. Along with the parent compounds, there are an enormous number of distinct lipid metabolites.

Lipids are a remarkably heterogeneous, diverse group of compounds related not by their chemical composition, but rather by physical properties, such as solubility in nonpolar solvents and insolubility in water. This physico-chemical classification indicates that there may not be obvious biological connections between various lipids. Precursor and derived members of lipid families include such divergent compound classes as fatty acids, steroids, sterols, and hormones, as well as more complex structures, such as phospholipids. Simple lipids are esters of fatty acids with alcohols, e.g., fats and waxes. Complex lipids include: (1) *phospholipids* that contain a phosphoric acid residue, often associated with a nitrogen-containing base and linked to a sphingosine or glycerol; (2) *glycolipids* that are formed when a carbohydrate is attached to a lipid that can range from a fatty acid to a sphingolipid or diglyceride; and (3) *lipoproteins* in which lipids are combined with a variety of proteins, usually in noncovalent assemblies, such as low-density and high-density lipoproteins (LDLs and HDLs).

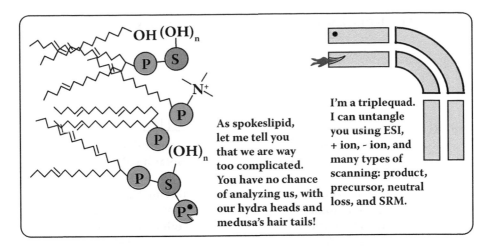

As spokeslipid, let me tell you that we are way too complicated. You have no chance of analyzing us, with our hydra heads and medusa's hair tails!

I'm a triplequad. I can untangle you using ESI, + ion, - ion, and many types of scanning: product, precursor, neutral loss, and SRM.

Mass spectrometry has become a major tool in lipidomics, because it is now possible to accrue and handle a vast amount of information that reflects the spatial and temporal alterations in the composition and quantities of lipid species in normal (physiological) and perturbed (pathological) states. The soft ionization techniques, primarily ESI and MALDI (and APCI for nonpolar lipids), are especially well suited for the simultaneous and rapid detection of entire classes of lipid families. The advent of nano-ESI combined with the multiple types of scans available using a triple quadrupole enabled the analysis of total lipid extracts from cells. Such extracts are usually obtained either by liquid:liquid or solid phase extractions. Examples of classes of lipids that can be analyzed in such extracts, using a combination of positive and negative ionization and precursor and neutral loss scanning, include sphingomyelins, phosphatidylcholines, phosphatidylinositols, phosphatidylinositolphosphates, phosphatidylethanolamines, phosphatidylserines, phosphatidylglycerols, phosphatidic acids, and their plasmalogen analogs. Types of negative ion CID MS/MS spectra obtained for phosphatidylinositols and phosphatidylserines are shown in Figure 3.26. The development of MALDI-based imaging techniques has added new levels of characterization because they facilitate the direct detection, identification, and quantification of different lipids or entire lipid families, such as phospholipids, together with their locations within individual tissues (Sections 3.6 and 4.13).

There are three major analytical categories in lipidomics: *untargeted*, *focused*, and *targeted*. The first one aims to analyze all lipids in a biological system without any preliminary information on the molecular or fragment ions, followed by sorting/grouping of them using pattern recognition techniques. This is a truly *global (shotgun)* approach that may not need initial chromatographic separation and is best accomplished using high-resolution techniques, such as FT-MS. Sophisticated computation and bioinformatics techniques have been developed to handle the vast quantities of data obtained from profile studies, particularly when the objective is to find and validate putative biomarkers. The immense structural diversity of stereoisomeric and regioisomeric lipid species remains a challenge with respect to understanding their metabolic pathways or diagnostic relevance.

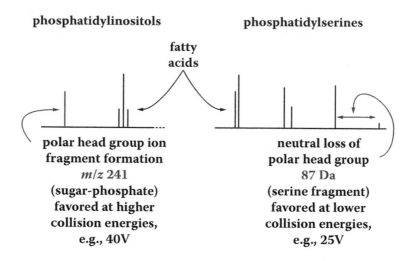

FIGURE 3.26   MS/MS product ion spectra of representative phospholipids.

The strategy in *focused lipidomics* is to utilize specific fragment ions, obtained via product, precursor, or neutral loss scanning (Section 3.3.3.1), to categorize all lipids in a particular class. Specificity can be improved with appropriate chromatographic separation, such as reversed- and normal- (hydrophilic liquid interaction chromatography (HILIC)) phase LC, as well as GC.

Targeted lipidomics, where quantification of specific lipids is the primary intent, is based on LC- or GC-MS/MS analysis, usually in selected reaction monitoring mode (Section 3.3.3.1). Quantification may be absolute or relative, with the former using conventional quantification strategies with internal standards (preferably isotopically labeled) or the standard addition method. Relative quantification is usually employed for comparison purposes, e.g., wild- and mutant-type lipid profiles. An alternative approach for relative quantification is based on isotope labeling through metabolism, i.e., introducing stable isotopes into the growth medium of cell cultures.

3.5   BIOPOLYMERS

The ability to work with biopolymers is probably the most spectacular of all the recent changes in mass spectrometry. Examples include the determination of the molecular masses of proteins along with the sequencing of their constituent amino acids and exploring the types and locations of post-translational modifications that occur. Other studies of proteins include investigating their three-dimensional structures, and their intercalations with small molecules and other proteins. Proteomics is by no means the only area of biopolymer omics activity, as there are also nucleic acids (genomics/nucleomics) research and the study of carbohydrates (glycomics), with the latter being particularly relevant in glycoprotein research.

3.5.1   PROTEOMICS

The first step in the generation of a protein is the synthesis of an amino acid chain derived from the sequence of the nucleotides in DNA. The sequence is encoded within a *gene*, a segment of DNA, which provides both the message for the order of the amino acids (sequence) and the mechanism by which the expression of this information is controlled. The term *genome* refers to all genetic information carried by a cell or organism, while *genomics* is the study of the structure and function of the genetic material in chromosomes. The spectacular success of genomics has provided complete DNA sequences for many organisms. However, it has been proven unequivocally that genomic analysis is insufficient to understand cellular function in both normal and diseased states. For example, the level of transcription of a gene provides only an estimate of the expression of the transcribed protein, as genetic analysis provides little or no information on the formation of *protein-protein complexes*, the potentially significant processes of *protein degradation*, and the all-important *post-translation modifications*, as most proteins are not active until they have been phosphorylated (Section 3.5.1.5) or glycosylated (Section 3.5.3). In the case of prokaryotes, transcripts may lead to more than one protein.

The term *proteome* refers to the total protein complement encoded within, and expressed from, a genome. *Proteomics* is the systematic study of those proteins, be they derived from the entire genome, from particular tissues or cells, or originating from healthy as well as diseased states. Proteomics has several, widely differing aims, including: (1) identification of proteins on a large scale, (2) identification of post-translational modifications (PTMs) (Section 3.5.1.5), (3) studies of protein-protein interactions, (4) comparison of the differential expression of specific proteins in healthy and diseased states, (5) exploration of the role(s) of specific proteins in metabolic pathways, and (6) development of drugs based on the presence/absence of specific proteins in diseases.

When the identity of an individual protein is known, i.e., if there is a known entry point, the proteomic strategy is *targeted*, and intended to explore the function and relationship of the target protein with small molecules or with other proteins. When there is no known entry point for the examination of the biological process or disease studied, the proteomic strategy is *global*, i.e., aims to analyze the entire protein complement during the course of changes in a cellular system. Such analyses involve the display and quantification of all proteins in the given system, and also the identification of those proteins in which there are quantitative changes.

Mass spectrometry is playing an ongoing and rapidly increasing role in the above aims. In the initial stages of proteomic research there was an understandable drive to identify as many proteins as possible. This was very valuable, as it led to the creation of large databases that can be searched using *peptide mass maps* (i.e., the list of peptide masses generated from proteolysis), or limited sequence data obtained by MS/MS. These libraries are important in enabling researchers to avoid "reinventing the wheel," i.e., undertaking *de novo* sequencing of already identified proteins.

3.5.1.1 Peptide Sequencing

The sequence of amino acids is unique to each protein. Determination of sequences enables the identification of specific proteins by comparison with information in databases or through *de novo* efforts to characterize unknown proteins. Sequencing with MS is accomplished using CID, or IRMPD in FT-ICRMS. These low-energy processes favor fragmentation at peptide bonds, in particular at the junction of the nitrogen and the carbon of the carbonyl (C=O) group. The resulting fragments are termed *b* and *y* ions (Figure 3.27). Those from the amino end of a chain are called *b* ions and are numbered per amino acid from that end of the chain. Ions numbered from the acidic end are called *y* ions. The positively charged [peptide-NH$_3$]$^+$ (*y*) ions are formed when two protons are added to the peptide–NH– fragment, which is formed by the direct cleavage of the peptide bond (–CO–NH–). Accordingly, the –NH$_3^+$ ion is technically a *y"* (*y* double-prime) ion. However, the use of *y"* is now uncommon, with fragments usually being referred to as *b* and *y*. Fragmentation at other locations around the peptide bond generates the *a* (*x*) and *c* (*z*) species. The *c* (*z*) fragments are important for the identification of phosphorylation sites.

   Although a sequence can be established for a protein, there are often other intense ions in the spectra that are not accounted for by the obtained sequence. Reasons for the additional ions include the formation of internal fragments, i.e., an interior portion of the amino acid chain. Additionally, the cyclization of the peptide ion and its subsequent fragmentation can release still other fragments that may include the amino acids that corresponded to the original acidic and amino termini.

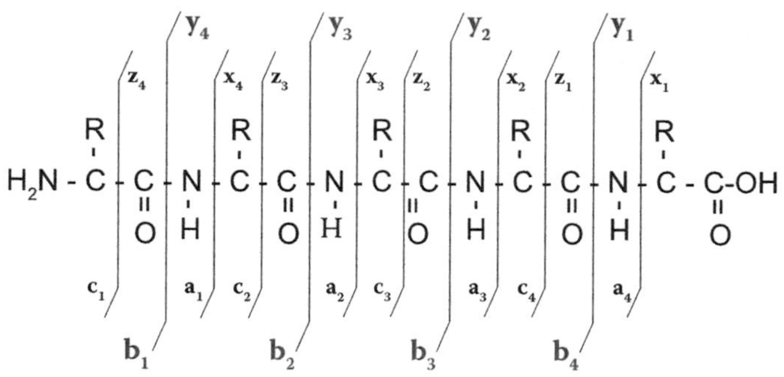

b and *y* ions are the fragment ions most commonly observed in low energy collision-induced dissociation (CID), the fragmentation process most frequently used in peptide sequencing.

y ions should be termed *y"* ions because of the two extra protons added, to the nitrogen during the fragmentation process, that generate the positive ion, [peptide-NH$_3$]$^+$. The term *y"* is falling into disuse.

Fragmentations that form *c* and *z* ions are important in the identification of phosphorylation sites.

FIGURE 3.27 Peptide fragmentation.

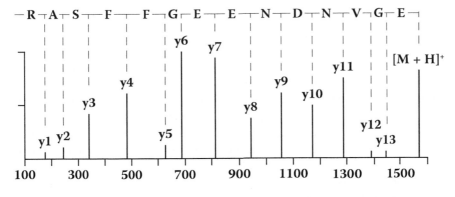

CID MS/MS favors formation of y (and some b) ions

FIGURE 3.28 Simplified spectrum of *y* ions showing the amino acid sequence of glutamine fibrinopeptide B.

An example of the *y* ions observed in peptide fragmentation, and in the sequence information obtainable, is shown in the simplified spectrum for glutamine-fibrino-peptide B (usually known as glu-fib) in Figure 3.28. The sequence is shown using single letter designations for the amino acids (see Table 3.8). The ions in the center of the spectrum (*y3* – *y11*) are more intense than those at the ends (*y1*, *y2*, *y12*, *y13*), which is typical in the CID spectra of most peptides, and can complicate the determination of sequences. Another frequently encountered effect in sequencing spectra is that the enhanced stability of the ion originating from fragmentation at a proline residue may make problematic the determination of the amino acid next to the proline. Neither case affects seriously the use of peptide sequencing in protein identification, as the data used to search the protein sequence databases are combinations of *partial sequences* and *molecular masses* of peptides.

That's one of my relatives who just ran into an inert gas and is all broken up .... I think he looks like a Picasso painting. However, he can be put back together sequentially and we will find out where he comes from.

3.5.1.2 Bottom-Up Protein Sequencing

The most common approach used in the identification of a protein is *bottom-up* sequencing (also called *shotgun proteomics*), where the protein is broken up into peptides and the molecular masses and sequences of the peptides are used to identify the protein. The first step is often to simplify a highly complex protein mixture using techniques such as one- or two-dimensional gel electrophoresis or immunoprecipitation. The resulting fractions are subjected to enzymatic (or chemical) degradation to generate peptides. Trypsin, the most frequently utilized proteolytic enzyme, cleaves proteins on the carboxyl side of arginine and lysine. The resulting peptides are referred to as *tryptic peptides*. There are several other endopeptidases/endoproteinases available to effect proteolysis at other locations in the peptide chain. The amount of enzyme used is generally kept low (e.g., ~1% of the amount of protein to be digested) to prevent autolysis of the protease and the appearance of undesired, enzyme-derived peptides. Proteolytic enzymes and their cleavage sites are listed in Table 3.7.

The masses and sequences of peptides, together with the locations of PTM, are determined with MS and MS/MS using ESI and MALDI, usually in combination with LC (Figure 3.29). If only MS is used, the set of molecular masses obtained for the peptides is called a *peptide mass map/fingerprint*. If the sample consists of a single protein, the peptide mass map can be matched to available protein databases to identify the protein. When there are multiple proteins in the sample, as is often the case, mass map–based attempts at identifications are inadequate, and MS/MS data on individual peptides are required to identify the proteins.

Accurate mass measurement is important in sequencing, e.g., to distinguish between the isobaric glutamine and lysine, both of which have nominal residue masses of 128 but different exact masses (128.0586 and 128.0950, respectively). Leucine and isoleucine are isomers and cannot be distinguished by CID. However,

---

TABLE 3.7
Cleavage Sites of Endopeptidases/Endoproteinases

Enzyme	Cleaves After (i.e., carboxyl side):	Except When Cleavage Would Occur Before:
Trypsin (pH 8)	K, R	P
Arg C (pH 8)	R	P
Asp N (pH 4–9)	Before D	
Chymotrypsin (pH 8)	F, L, W, Y	P, after PY
Glu C (pH 7.8)	E	P, E
Glu C (pH 4)	D, E	D, E
Lys C (pH 8)	K	
Pepsin (pH 2–4)	F, L	
Proteinase K (pH 6–10)	C, F, G, M, S, W, Y	
Cyanogen bromide[a]	M	

[a] Cyanogen bromide treatment is a chemical process that cleaves only at methionine residues.

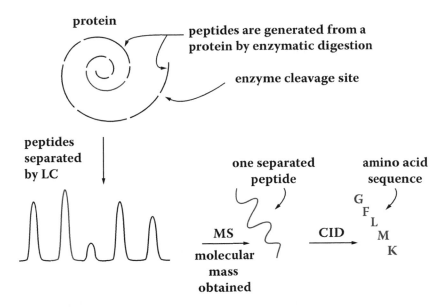

In the bottom-up strategy the protein is first digested enzymatically.
LC is used to separate the resulting peptides and introduce
  them into the mass spectrometer.
MS and MS/MS are used to obtain the masses and determine the
  sequences of the peptides.

FIGURE 3.29   Bottom-up protein sequencing.

correct assignments can be made for these two amino acids when the databases
used also contain genomic data. The amino acids are generally referred to by using
their *residue* masses, i.e., the masses after the loss of water during the formation of
peptide bonds. The water at the ends of the peptide, i.e., in the NH_2 and COOH, and
also any PTMs have to be factored into sequence determinations. The masses of the
residues of the 20 biologically important amino acids are listed in Table 3.8.

The processing of spectra for sequence information and the interrogation of
databases (accessed via the Internet) for protein identification have become highly
sophisticated parts of protein mass spectrometry. The unique sequences of proteins,
combined with the large amount of sequence information in the databases, have
the combined advantage that relatively little data are required on the enzymatically
derived peptides to identify a protein. Determining the molecular masses and partial
sequences for two or three peptides is usually sufficient to identify a protein.

There are a number of pitfalls in the bottom-up approach to protein analysis:
(1) Incomplete enzymatic digestion can result in longer than expected peptides,
(2) Only 50–70% (frequently much less) of the peptides from a given protein are
usually identified, (3) Although CID spectra do provide the sequence, other intense
ions are often present that cannot be accounted for (they may originate from internal

TABLE 3.8
Residues of Amino Acids

Amino Acid Residues		Residue Formula	Monoisotopic Mass (Da)	Average Mass (Da)
Glycine	G	C_2H_3NO	57.0215	57.0513
Alanine	A	C_3H_5NO	71.0371	71.0779
Serine	S	$C_3H_5NO_2$	87.0320	87.0073
Proline	P	C_5H_7NO	97.0528	97.1152
Valine	V	C_5H_9NO	99.0684	99.1311
Threonine	T	$C_4H_7NO_2$	101.0477	101.1039
Cysteine	C	C_3H_5NOS	103.0092	103.1429
Leucine	L	$C_6H_{11}NO$	113.0841	113.1576
Isoleucine	I	$C_6H_{11}NO$	113.0841	113.1576
Asparagine	N	$C_4H_6N_2O_2$	114.0429	114.1026
Aspartic acid	D	$C_4H_5NO_3$	115.0269	115.0874
Glutamine	Q	$C_5H_8N_2O_2$	128.0586	128.1292
Lysine	K	$C_6H_{12}N_2O$	128.0950	128.1723
Glutamic acid	E	$C_5H_7NO_3$	129.0426	129.1140
Methionine	M	C_5H_9NOS	131.0405	131.1961
Histidine	H	$C_6H_7N_3O$	137.0589	137.1393
Phenylalanine	F	C_9H_9NO	147.0684	147.1739
Arginine	R	$C_6H_{12}N_4O$	156.1011	156.1859
Tyrosine	Y	$C_9H_9NO_2$	163.0633	163.1733
Tryptophan	W	$C_{11}H_{10}N_2O$	186.0793	186.2099

cleavages or are the products of cyclization and alternative fragmentation) and are disconcerting, if only because of their prominence, (4) There is lack of information about PTM, particularly phosphates, because of their loss during CID, and (5) While ETD/ECD can provide the locations of phosphates, the technique is often effective only on larger, multiply charged peptides that are hard to sequence and frequently require the use of enzymes other than trypsin to generate the larger peptides that are necessary. Despite these disadvantages, bottom-up sequencing has been very successful and is by far the most used method. In addition, the bottom-up approach has generated the large databases that are searched to identify proteins (Section 3.7).

3.5.1.3 Top-Down Protein Sequencing

Top-down sequencing, an alternative methodology for protein characterization, is based on introducing the *intact* protein, *without* proteolytic digestion, into a tandem mass spectrometer that has an ESI source, and progressively removing amino acid residues from the peptide chain (Figure 3.30). The resolving power and mass

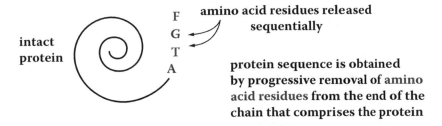

intact protein

F
G ← amino acid residues released
T ← sequentially
A

amino acid residues released sequentially

protein sequence is obtained by progressive removal of amino acid residues from the end of the chain that comprises the protein

In the top-down method proteins are introduced into the mass spectrometer intact, i.e., without proteolysis.

Individual amino acid residues are released from the end of the peptide chain using IRMPD or ECD.

The spectra generated are complex as the removal of each amino acid results in another set of multiply charged ions. The number of ^{13}C atoms in the ions must be determined because of the effect this isotope has on the observed molecular masses.

FT-ICRMS is the most effective instrument because the resolution available facilitates spectrum interpretation by enabling charge state determination on highly charged ions.

FIGURE 3.30   Top-down protein sequencing.

accuracy of FT-ICRMS instruments are key capabilities for the production and interpretation of top-down spectra.

This method has the advantage that there are no issues associated with peptide generation and separation. Furthermore, the MS/MS process can be adjusted to preserve post-translational modifications, e.g., by switching from CID/IRMPD to ETD/ECD. Difficulties include the large quantity of *purified* protein required (~20 times more than for bottom-up sequencing and purity is not an issue because LC separation is part of the protocol) and the complexity of the data because all the ions are multiply charged. Another potential problem is the ^{13}C component of proteins, as the molecular mass observed for intact species includes an unknown number of ^{13}C atoms. It is difficult to determine the exact number of ^{13}C atoms but accurate and precise isotope ratio measurement is helpful. Furthermore, the remaining peptide chain, after the removal of the individual amino acid residue, is a new species with its own multiply charged ion envelope, thus there is the likelihood of peaks from the different envelopes overlapping and interfering with one another. The resolution of FT-ICRMS is essential for elucidation of the different charge envelopes.

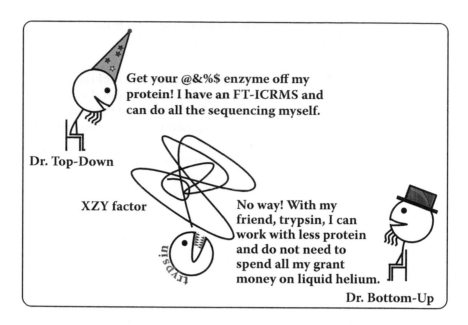

3.5.1.4 *De Novo* Protein Sequencing

When information for a protein cannot be found in databases, the entire amino acid sequence must be derived from mass spectra. A similar situation can occur when there is extensive PTM that interferes with automated data processing.

In bottom-up sequencing, with trypsin as the proteolytic enzyme, it is usually only possible to observe peptides that represent up to 50–70% of the protein; thus, only a fraction of the composition is obtained. Furthermore, it is not possible to arrange the peptides in any specific order, so additional digestions with different enzymes are required to generate complementary sets of data.

Additional techniques can provide further information when the structure of a peptide is unknown. One method is to carry out the protein digestion in water that is labeled with 50% ^{18}O, yielding an isotopically labeled ($^{16}O/^{18}O$) doublet for every cleaved peptide. When such peptides are fragmented the *y* ions (and other C-terminal ions) can be readily identified, as each will retain the $^{16}O/^{18}O$ doublet, whereas the N-terminal fragments will have normal isotope distributions that contain only the ^{16}O isotope.

The major steps in *de novo* protein sequencing are: (1) digest multiple samples of the unknown with several enzymes to produce multiple sets of peptides, (2) analyze these sets by MS to determine their *m/z* ratios, (3) use MS/MS to obtain the amino acid sequence of each peptide, and (4) cross-reference the sequenced chains, cleaved by the different enzymes, to derive the overall sequence of the protein (Figure 3.31).

De novo sequencing is not a trivial undertaking. Although automatic algorithms have been developed, manual interpretation of both the CID spectra and the alignment of sequences improves the quality of the data for each series acquired using the different enzymes. The need for *de novo* sequencing is diminishing as more

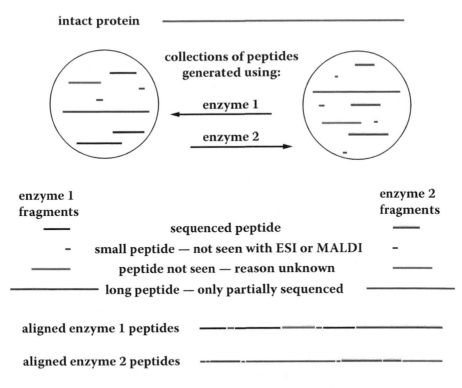

FIGURE 3.31 *De novo* protein sequencing.

genomes are sequenced. The complementary information provided by partial amino acid sequence and genomic data can be used to provide tentative identification of unknown proteins. Genomic data can also be used to distinguish between the isomeric amino acids leucine and isoleucine.

3.5.1.5 Post-Translational Modification (PTM)

The functionality of proteins is often brought about by *post-translational modification* (PTM) of the peptide chain. The concept of PTM includes all covalent changes not encoded for by the gene, with the changes occurring to almost any part of a protein. Newly synthesized polypeptide chains acquire their three-dimensional structures spontaneously; to attain their functional form they may also have to undergo PTM through covalent chemical reactions. There are more than 200 known PTMs, including acetylation, methylation, farnesylation, glycosylation, and most importantly, phosphorylation. Because PTM involves the addition (or loss) of mass to

(from) the peptide chain, mass spectrometry is well suited to detect, identify, local-
ize, and quantify modifications, often at the femtomole level, for endopeptidase-
generated peptides up to about 30 residues.

Proteins are converted into *phosphoproteins* when a negatively charged phosphate
group is added, via an oxygen linkage, to the hydroxyl group of serine, threonine,
or tyrosine. Phosphorylation acts as a *molecular switch* controlling when a protein
is performing a biological task that may include regulating intracellular functions
and metabolic pathways. Phosphorylation is among the most frequent PTM, with
approximately 30% of proteins cycling between phosphorylated and dephosphorylated forms, mediated by phosphotransferase enzymes. The phosphate groups are
added by *protein kinases*, while phosphate removal is controlled by *phosphatases*.

Mass spectrometry is an excellent technique to search for phosphates because a
phosphoryl group ($-PO_3H$) provides an 80 Da mass increment; i.e., if a peptide has a
mass 80 Da higher than expected, then there is a likelihood that it is phosphorylated.
Determining the location of the phosphate group on a peptide chain is problematic
with CID-MS/MS because the phosphate groups are labile and are often lost during
the formation of b and y ions. One approach to solve this problem is the use of fragmentation reactions that favor the formation of c and z ions, a process that preserves
the phosphate groups (Figure 3.32). Two strategies for the identification and location
of phosphate groups are electron transfer dissociation (ETD) and electron capture dissociation (ECD) (Section 3.3.2.3). In ETD a multiply charged (usually triply charged)
peptide cation is reacted with a singly charged anion, resulting in the transfer of an
electron to the peptide. The equilibration of this added energy leads to fragmentations
that favor the formation of c and z ions that allow the location of the modification. ETD,

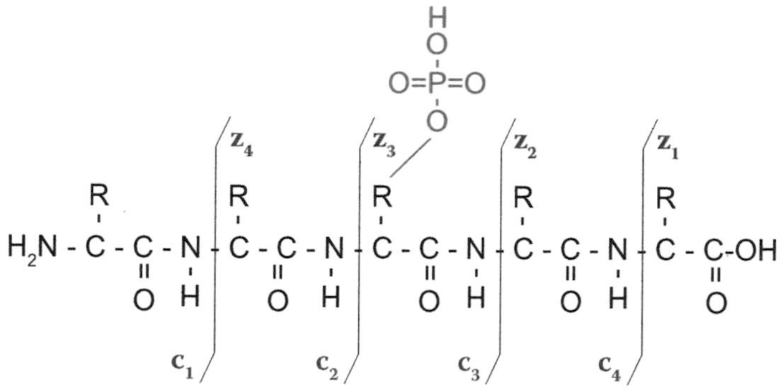

**c and z ions are the most commonly observed fragment ions in
the mass spectra obtained for peptides after electron transfer
dissociation (ETD) or electron capture dissociation (ECD).
ETD and ECD preserve post-translational modifications (PTM),
in particular phosphorylation, but still provide the peptide
sequence and, therefore, allow determination of the location
of the PTM.**

FIGURE 3.32   ETD/ECD peptide fragmentation.

used on LIT-orbitrap and QTOF type MS/MS systems, takes place in the CID cell. In the ECD process, used on FT-ICRMS instruments, the required low-energy electrons, obtained from a filament, are allowed to react directly with the peptide ion within the ICR cell. The spectra produced are similar to those obtained by ETD.

In addition to the naturally occurring PTM, there may be alterations consequent to sample preparation, including those added intentionally, such as the reduction and alkylation of cysteines, or unintentionally, e.g., the oxidation of methionine.

3.5.1.6   Some Other Ions to Know About and Avoid

When analyzing peptides by LC-MS, especially with nano-LC, several of the contaminants listed earlier (Section 3.2.6) are encountered, particularly those that are PEG-based and various siliconates. There are also contaminants specific to peptide analysis, such as the products of autolysis, that occur when too much trypsin is added (or when the sample contains less protein than expected) (Table 3.9a). As a rule of thumb, trypsin should be present at a concentration that is ~1% of the protein to be digested. Another common contaminant is keratin that is introduced by incorrect handling of samples or by accidental contamination during sample collection (Table 3.9b).

3.5.1.7   Protein Conformation

Conformation refers to the spatial arrangement of the atoms in a molecule, e.g., the three-dimensional (3-D) shape of a protein. Proteins roll off the protein-making

---

TABLE 3.9
Major Porcine Trypsin (a) and Human Keratin (b) Ions

(a) Porcine Trypsin

Peptide (Da)	Major Charge State	*m/z*	Sequence	Modifications
841.50	1	841.50	VATVSLPR	
1,044.56	2	522.28	LSSPATLNSR	
1,062.50	2	531.25	APVLSDSSCK	Carbamidomethyl C
2,210.10	3	737.71	LGEHNIDVLEGNEQFINAAK	
2,282.17	3	761.73	IITHPNFNGNTLDNDIMLIK	
2,298.17	3	767.06	IITHPNFNGNTLDNDIM*LIK	Oxidation M

(b) Human Keratin

Peptide (Da)	Major Charge State	*m/z*	Sequence	Type of Keratin
826.45	2	414.24	PEIQNVK	k2
972.52	2	487.27	IEISELNR	k2
1,030.59	2	516.30	VLDELTLTK	k1
1,063.60	2	532.81	LASYLDKVR	k1
1,474.74	2	738.38	WELLQQVDTSTR	k2
1,474.78	3	492.60	FLEQQNQVLQTK	k2
2,500.24	3	834.42	SKAEAESLYQSKYEELQITAGR	k2
2,509.12	3	837.38	EIETYHNLLEGGQEDFESSGAGK	k1

ribosome assembly line as linear strings of amino acids. To acquire biological functionality, a protein must first fold into a unique 3-D structure that is inherent to the sequence of its amino acids. In eukaryotes, proteins achieve these native states in seconds-to-minutes after their release from the ribosome. While these conformations are normally long-lived, they are also malleable and can change when the protein interacts with other proteins or small molecules inside a cell, or undergoes PTM. The structural changes induced by PTM, protein:protein and protein:small molecule interactions (all of which are generally reversible) are usually critical to the normal functioning of proteins.

Proteins must be in the vapor phase and ionized for mass spectrometric analysis. There have been long-standing discussions as to whether the structure of vaporized and ionized proteins bears any resemblance to the form in which they exist when in their biologically active states. It is certainly possible to alter the conformation of a protein ionized using ESI by changing the composition of the injected solution, by altering the voltages in the source, or by varying the pressure in the step-down region behind the inlet cone in the ion source.

The accessibility of the primary amino groups of proteins affects the number of charges that can be added to the molecule during ionization and suggests the degree to which a protein (ion) is folded. Many amino groups are concealed within the folds of proteins, while in their native configurations, and these are unavailable for ESI. Accordingly, the closer a protein is to its native structure the lower will be the number of charges on the ions observed. When a protein is *denatured*, e.g., by exposure to an organic solvent or by a change of pH, the noncovalent interactions necessary to hold the folded chain together are disrupted and the protein begins to denature. The unfolding leads to an increase in the number of sites that can be protonated (Figure 3.33).

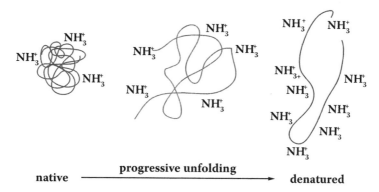

In electrospray ionization the number of charges carried by a
  protein depends on the accessability of the amino groups.
As a protein denatures the number of exposed amino groups
  increases, consequently the number of charges on the ionized protein
  also increases.
The increase in the number of charges moves the ion envelope
  from higher to lower *m/z*.

FIGURE 3.33   Effects of the conformational state on the number of charges carried by a protein.

The importance of the effects of denaturation on protein structure, and on the mass spectra obtained, is illustrated by comparing the spectra of native and denatured horse heart myoglobin, a relatively small (oxygen-carrying) protein that contains a noncovalently bound heme group that is key to its biological activity. When dissolved in water the protein acquires 7 to 12 charges during ESI. Deconvolution of the data reveals that the protein is present in its holo form with a molecular mass of 17,566 Da; i.e., the heme group is retained within its 3-D structure (Figure 3.34).

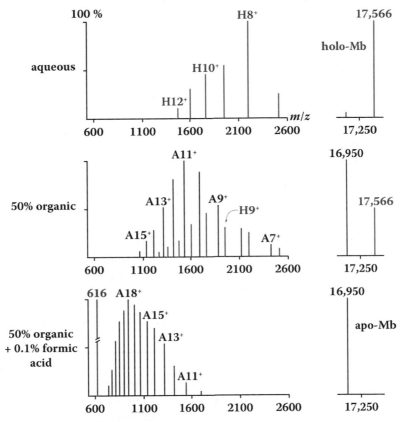

Myoglobin remains in the holo form in aqueous solution, with the non-covalently bound heme group present (MW 17,566).

Increasing the amounts of the organic solvent (methanol or acetonitrile) and the acidic components of the solution denatures the protein, enabling the acquisition of extra charges.

As the protein unfolds and carries additional charges the ion envelope moves to lower m/z.

The unfolded protein releases the heme group giving an intense ion at m/z 616 and multiply charged ions that correspond only to apo-myoglobin (MW 16,950).

FIGURE 3.34 Mass spectra and deconvoluted masses for ESI analysis of horse heart myoglobin using different solvents and acidities.

When an organic solvent, e.g., acetonitrile, is added to the solution prior to ionization, the number of charges increases; subsequent deconvolution of the spectrum shows both the holo and apo forms of the protein (17,566 and 16,950 Da, respectively), suggesting that the protein is beginning to unfold, leading to the loss of the heme group. When some acid (0.1% acetic or formic) is added to the solution, the protein is fully denatured and the spectrum has 10 to 23 charges. There is also an abundant ion at m/z 616 for the heme group. Deconvolution shows that the molecular mass is 16,950 Da, corresponding to the apo form (Figure 3.34).

Obtaining spectra in both the native and denatured forms of proteins is important for understanding their 3-D structures. As illustrated by myoglobin, an important aspect of protein conformation is whether a protein is capable of holding, noncovalently, a small molecule within its structure. A change in the molecular mass between native and denatured states indicates the presence of a noncovalent species. Given that the capture of a small molecule can significantly alter the activity of a protein, there are obvious pharmaceutical implications for obtaining such information.

It is difficult to obtain spectra of proteins in their native state because the protein has to be in an aqueous solution. In ESI, droplets are formed and then dried to release ions. When the solution is 100% aqueous, the ESI process is disrupted by the surface tension of water that tends to prevent the orderly disintegration of the droplets; this reduces sensitivity. To maintain the natural configuration, the solution of the analyte may need to be buffered; however, all buffers contain salts that can ionize preferentially, yielding ions that can dominate the spectra. When possible, ammonium acetate should be used as the buffer, as the ammonium ions do not interfere with the formation of multiply protonated protein ions. The reduced charge states of native structures result in increased m/z ratios that may limit the types of mass analyzer that can be used. Figure 3.34 illustrates differences in the required mass range as the charge envelope changes from having a mode at $[M + 8H]^{8+}$ (m/z 2,200) for the native structure to a mode of $[M + 22H]^{22+}$ (m/z 943) for the denatured myoglobin.

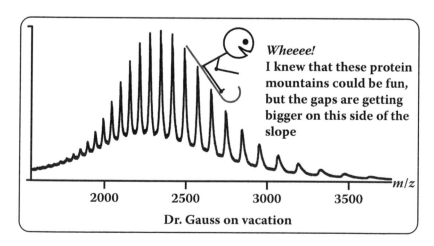

Dr. Gauss on vacation

Determining the intact molecular mass of a protein in both its native and denatured states can provide additional information, assuming that the amino acid sequence is

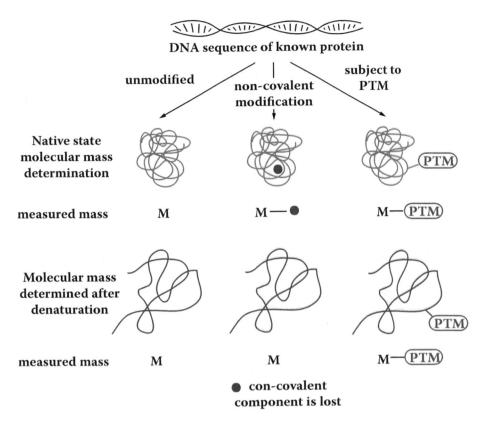

DNA sequence of known protein

unmodified — non-covalent modification — subject to PTM

Native state molecular mass determination

measured mass   M   M—●   M—(PTM)

Molecular mass determined after denaturation

measured mass   M   M   M—(PTM)

● con-covalent component is lost

Determining the molecular mass of a known protein in both native state and after denaturation shows whether the protein has been modified and if so whether the modification is non-covalent or covalent.

FIGURE 3.35 Mass determinations for proteins with noncovalent and covalent modifications.

known. When the molecular mass of the native state is higher than that of the denatured configuration, then a noncovalent component must be present, as illustrated above for the heme group and schematically in Figure 3.35. If, however, the mass of the denatured protein is higher than that predicted by the sequence, then the additional mass is covalently bound and provides evidence of a PTM. This can alter the strategy employed to obtain the identity and location of the modification when sequencing the peptides, e.g., using ETD instead on CID when the presence of a phosphate is suspected.

3.5.1.8 Hydrogen/Deuterium Exchange (HDX)

Elucidation of the dynamics of protein conformation and the contact surfaces in protein interactions with other proteins, nucleic acids, etc., is important to the understanding of protein function. *Hydrogen/deuterium exchange* (HDX, H/D, H-D) is a mass spectrometric strategy to obtain such information.

When a proton that is attached to a heteroatom (e.g., –NH, –OH, –SH) is exposed to deuterium, usually from D_2O, exchange can take place between the protons and deuterons yielding –ND, –OD, and –SD. Subsequent exposure of these deuterons to protons results in *back exchange*. The rate of the back exchange depends on the location of the heteroatom. The exchange for proteins is rapid when the deuteron is located on an amino acid side chain; the kinetics cannot be followed by mass spectrometry. At the other extreme, even if C–H bonds do undergo H-to-D exchange, it is not on an experimentally relevant timescale. When the protons in the peptide bonds (–NHCO–), which form the backbones of polypeptides, are exchanged with deuterium (to give –NDCO–), the extent of the D incorporated in the peptide chain can be obtained using LC-MS. In this case, the back exchange to the proteo form occurs on a timescale that permits mass spectrometric analysis, provided that the analyses are conducted at pH 2.6 and at 0 °C. Any back exchange must be accounted for during data evaluation because orders of magnitude differences in the rates of exchange can be induced when deuterons are located at different sites within the 3-D structure of the protein. Influencing factors include whether (1) bonds are buried within the hydrophobic core, (2) the atoms are participating in the hydrogen bonding of the α-helices and β-sheets or are protected from exposure by the higher-order structures of folded proteins, and (3) there is interaction of structural components with other proteins or small molecules.

HDX studies can be carried out in an "exchange in" mode, i.e., when the first step is labeling of a proteo form of a protein with deuterons, or in an "exchange out" mode, i.e., when a deuterated protein is labeled with protons. The former method is used more frequently because it does not require that the initial deuteration be complete. Deuterons can be introduced into the protein (or protein complex) either at physiological pH or, when more extensive labeling is required, by pH pulsing where the pH is briefly increased to ~10 to create "reversible" denaturation, during which the protons of interior amides are made available for exchange. The samples must be chilled to 0 °C immediately after the exchange, and the pH adjusted to 2.6.

The proteins are digested with pepsin (which is active at pH 2.6 and 0 °C) and then analyzed by LC/MS to determine the extent of deuteration and for the calculation of kinetic data. Back exchange still occurs even at pH 2.6 and 0 °C, so the enzymatic digestion and chromatography have to be carried out rapidly, in 10–20 min. Alternatively, instead of digesting the proteins, it is possible to determine the overall extent of deuterium incorporation by using ESI-MS to measure the molecular mass of the intact protein.

HDX can be used in various ways to obtain information on certain structural aspects of proteins. In the case of an intact protein, where physiological conditions may affect the conformational folding, the extent to which deuterium is incorporated into the overall molecular mass, as conditions are altered, enables the determination of the extent and rates of folding. Investigating the rates of exchange occurring, as a protein is exposed to D_2O over a time course, provides information on the locations and interactions of amino acids. However, care must be taken to understand the effects of the substructures, such as β-sheets and α-helices, on the exchange rates as well as on the normal D-to-H back exchange.

Another application of HDX is to elucidate which amino acids form the contacting faces of two interacting proteins. In such experiments the first step is to conduct HDX on each protein independently. The deuteration should be brief (without pH pulsing), so that the labeled peptide bonds are those on the surface of the protein. Digestion and LC-MS are then used to determine which peptides are deuterated and constitute the surfaces of each protein. In a second experiment, the two unlabeled proteins are allowed to interact, so that those amino acids that form the interface will no longer be from exposed surfaces. The protein complex is then subjected to HDX during which the peptides that have now formed the interface will not be labeled. After digestion and LC-MS, a comparison of the data from the HDX conducted after complex formation with that from the individual protein labeling experiments will make it possible to identify those peptides containing peptide bonds that are not deuterated, and therefore the amino acids that form the protein:protein interface (Figure 3.36).

Difficulties of HDX experiments result from the need to control and understand the biochemistry and physical-chemical processes of the exchange of deuterons on and off the amide bonds. Both processes occur before the samples are introduced into the mass spectrometer.

3.5.1.9 Quantification of Proteins

Mass spectrometry-based techniques for the quantification of proteins have gained increasing popularity over the past decade, particularly to assess the differences between two (or more) physiological states of biological systems, e.g., normal and diseased states. Advantages of the MS-based quantification techniques include sensitivity, accuracy, and importantly, automated high-throughput operation. Because of significant differences in the efficiency of ESI or MALDI, the signal intensities from peptide ions are generally not used for direct (absolute) quantification, but rather for relative quantification. Other factors making protein quantification by MS difficult include the large number of different proteins in tissues and the substantial variations in their molecular mass and concentrations. Blood plasma is particularly confounding because the 10^6–10^9-fold excess albumin concentration over other proteins makes the quantification of those proteins difficult.

There are two significantly different strategies for quantitative proteomics using mass spectrometry: *stable isotope labeling* (isotopic tagging) and *label-free* methods. Stable isotope labeling is used to create a specific *mass tag*. This strategy is based on (1) stable isotope-labeled peptides being *chemically identical* to their native counterparts and, consequently, having identical chromatographic and mass spectrometric properties, and (2) the mass differences between labeled and unlabeled forms of the analytes can be readily determined. Accordingly, quantification is possible by comparing the signal intensities from differentially labeled peptides.

The isotopic mass tags may be introduced into the proteins or peptides at different stages of the sample preparation by metabolic (*in vivo*) or chemical (*in vitro*) means. The preferred isotopic labels are ^{13}C and ^{15}N, as they do not induce changes in the LC retention times of the labeled species when compared to the retention of the same peptide with natural isotopic abundances. Variations in LC retention times can occur when deuterium labeling is used, making the interpretation of data difficult.

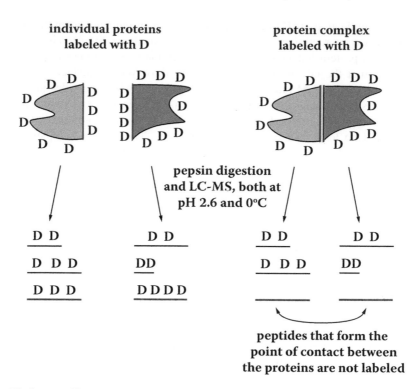

Hydrogen/Deuterium exchange (HDX) can be used to determine the
   peptides that are at the points of contact in protein complexes.
Labeling is conducted so that exchange only occurs
   for peptide bonds on the outer surfaces of the proteins.
Experiment 1: the unlabeled proteins are labeled individually to
   determine the surface amino acids.
Experiment 2: the unlabeled proteins are allowed to interact and
   then labeled. The peptide bonds at the interacting surfaces will
   not be labeled in this experiment so the amino acids that form
   these surfaces can be identified.
Digestion with pepsin and chromatography (0 °C and pH 2.6)
   followed by MS, are used to determine the locations
   of the deuterons.

FIGURE 3.36   A hydrogen/deuterium exchange (HDX) experiment.

In quantitative proteomics it is usual to talk in terms of moles (often picomoles),
rather than mass (weight), for the amounts of individual analytes. Conversion between
moles and mass is confusing. The following equation can help at the pmol level: pmol/
µg = 1,000/MW in kDa. Thus, for a 10 kDa protein, 1 µg ≡ 100 pmol, while for a
150 kDa protein, 1 µg ≡ 6.7 pmol. When a protein is digested, the same number of
moles of each peptide is produced as there were of the original protein. That is, for

10 pmol of protein there will be 10 pmol of each peptide, but the mass of the individual peptides (usually ng to μg) will be a function of their empirical formulae.

**So now it is perfectly clear how to
convert between mass (weight) and moles!**

3.5.1.9.1   Quantification Using Stable Isotope Labeling by Amino Acids in Cell Culture (SILAC)

The earliest possible point to introduce a stable isotope into a protein is *in vivo*, during the growth and division of the cells. *Stable isotope labeling by amino acids in cell culture* (SILAC) is a strategy where specific ^{13}C- or ^{15}N-labeled amino acids are added to the growth medium for one of a pair of cultures (Figure 3.37). For example, if arginine labeled with six ^{13}C atoms is used then the growing cells will incorporate the heavy arginine into all proteins that are being synthesized, and the molecular masses of the expressed proteins will be increased by 6 Da for every labeled arginine present. The increase in the molar mass of the labeled SILAC amino acids has no apparent effect on the growth rates or the morphology of the cells.

Given that organisms respond to environmental and physiological stresses, e.g., heat, specific nutrient deprivation, and toxins, it is assumed that there will be consequent changes in the concentrations of individual proteins. SILAC experiments elucidate which proteins are altered during cell growth by subjecting one of the two cultures to a physiological stress. The labeled and unlabeled samples are pooled at the end of the growth period and processed using standard protein recovery and digestion methods, followed by LC-MS and LC-MS/MS. Unlabeled (light) peptides and their labeled (heavy) counterparts elute from the LC at the same time, and their relative amounts can be determined from the ratios of their respective peak intensities. Identification of the proteins is obtained by MS/MS of the light peptides and database searching. The relative amount of each protein provides information on

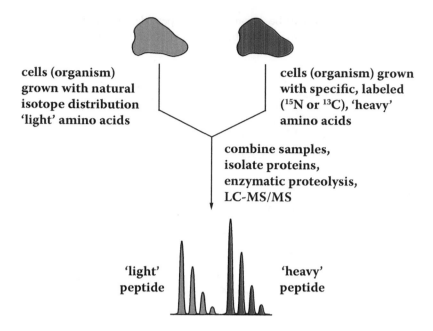

SILAC (stable isotope labeling by amino acids in cell culture) is an
    ***in vivo* labeling technique.**
Two identical cell cultures are prepared with the exception that one
    is grown in a medium containing specific, labeled (^{15}N or ^{13}C)
    'heavy' amino acids, e.g. $^{13}C_6$ arginine.
One of the cultures is subjected to physiological conditions that may
    alter protein synthesis, e.g., heat, lack of a key nutrient.
Changes in the concentrations of proteins are determined, using
    LC-MS, by the relative amounts of the 'light' or 'heavy' peptides
    after the samples have been combined, and the proteins isolated and
    subjected to enzymatic proteolysis.
Identities of the proteins are obtained from LC-MS/MS data on the
    'light' peptides.

FIGURE 3.37   SILAC methodology for *in vivo,* relative protein quantification.

the up- and down-regulation of the proteins under the altered physiological conditions, i.e., whether the concentrations of individual proteins change in response to
the stressor to which the cells were subjected.

SILAC is arguably the most accurate method for the quantification of individual
proteins. The technique is rapid and completed after six to eight cell doublings. The
strategy is particularly useful for the quantitative study of PTM residues, such as the
phosphorylation of tyrosine. The method is relatively inexpensive and convenient.
On the negative side, metabolic labeling of complex body fluids and higher organisms *in vivo* is difficult. Another limiting factor is the availability of suitably labeled
cell culture media.

3.5.1.9.2   Quantification Using Isotope-Coded Affinity Tag (ICAT)

Other approaches to isotope label-based quantification rely on adding the label to the proteins during the isolation process. One technique, *isotope-coded affinity tag* (ICAT), is based on using chemical probes that consist of three components: (1) an iodoacetamide-based *reactive group* that is specific for thiols, i.e., cysteines, (2) an ethylene glycol-based *linker* with either normal isotope distribution or containing a stable isotope label that can be differentiated by mass spectrometry, and (3) an *affinity tag* (biotin) used to isolate the labeled peptides. The heavy form of the linker is labeled with eight deuterium atoms or nine ^{13}C atoms. The light linker has the natural abundance of D and ^{13}C, and thus has mass differences of 8 or 9 Da less than that of the heavy chain. The linker is coupled to biotin with a cleavable bond.

When two proteomes (sets of proteins from cells, tissues, etc.) are to be compared quantitatively, one sample is labeled with an *isotopically light* probe (i.e., natural isotope abundance), while the other is labeled with an *isotopically heavy* (D_8 or $^{13}C_9$) version. The two samples are then combined (to minimize the effects of sample preparation) and digested with a protease. Next, the peptides labeled with the tagging reagents are isolated using a streptavidin column. The recovered peptides (including the linker) are released from the biotin with trifluoroacetic acid and analyzed using LC-MS and LC-MS/MS. The relative quantities of the proteins can be determined from the ratios of signal intensities of the differentially mass-tagged light and heavy peptide pairs (Figure 3.38). The proteins are identified by MS/MS of the light (D_0 or ^{12}C) peptides using database searches.

3.5.1.9.3   Isobaric Tag for Relative and Absolute Quantification (iTRAQ)

The technique of *isobaric tag for relative and absolute quantification* (iTRAQ) is based on covalently labeling the amines at N-termini and the side chains of peptides with isotope-coded tags after the proteins have been isolated form biological systems, denatured, and enzymatically digested. Here the isotope labeling is carried out at a much later stage than in both SILAC and ICAT. iTRAQ is an MS/MS technique that uses CID-derived ions for both the identification and quantification of proteins.

The method utilizes the reaction of N-hydroxysuccinimide (NHS) with primary amines. Two different sets of tags, 4-plex and 8-plex, i.e., with four or eight sets of reagents, are available, and the intention is to label all peptides in the sample with a specific, isotopically labeled marker (Figure 3.39). The 4-plex iTRAQ reagents consist of *reporter* groups that vary but are specific to each of the four reagents and yield characteristic ions upon MS/MS (*m/z* 114–117), a *mass balance group* (18–21 Da), and a *peptide reactive group* that is an NHS ester. The total mass of the isobaric tag is 145 Da for each reagent. The 145 Da unit is obtained by varying the stable ion composition of the reporter and mass balance groups, e.g., $114 + 21 = 145$ and $116 + 19 = 145$ (Figure 3.39). The four reagents, named after the masses of the reporter groups (i.e., 114, 115, 116, and 117), allow the simultaneous analysis of four samples. Thus, the derivatization reagents can be used to provide relative quantification of proteins under four experimental conditions, e.g., for genetically modified cell lines. The labeled ion released during MS/MS is called a *reporter ion* because it is a surrogate for the protein/peptide it represents.

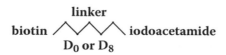

Example of ICAT reagent used to label cysteines.

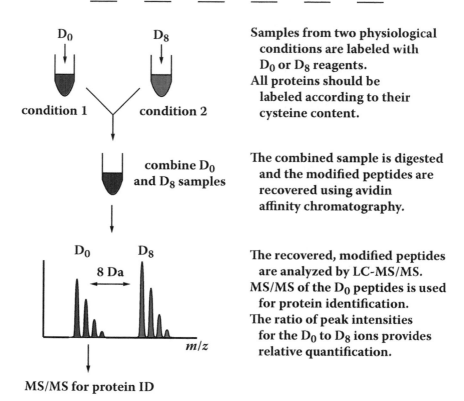

FIGURE 3.38   ICAT procedure for labeling cysteines.

The isobaric tag in the 8-plex reagent set has a mass of 305 Da, with reporter and balance components that vary between 113–121 Da and 184–192 Da, respectively. The 120/185 Da combination is not used because of the interference from the immonium ion of phenylalanine at m/z 120.

The labeling of peptides is carried out *in vitro* after proteolysis. The series of steps that make up an iTRAQ (4-plex) analysis are shown in Figure 3.40. The group transferred onto the peptide is composed of an N-methyl piperazine linked to a carbonyl group with differential labeling of each group using ^{13}C, ^{15}N, and ^{18}O to produce four different labels that add up to 145 Da (Figure 3.39). The individually labeled reagents are reacted with each set of peptides derived from four parallel experiments or four samples from a time course. The products of the labeling are then mixed and analyzed by LC-ESI-MS/MS or by MALDI-TOF/TOF with prior LC separation and robotic spotting of the LC eluate onto the MALDI plate. The N-methyl piperazine reporter

4-plex iTRAQ

145 Da

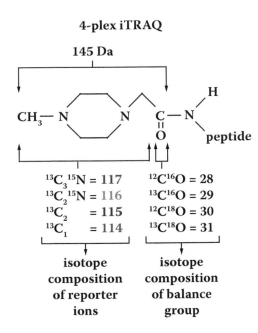

8-plex iTRAQ

The mass of the reporter/
balance group is 305 Da.
The larger balance group has
8 possible isotope formats
(184–192 Da except 185).
The reporter ions have the
same structure as the 4-plex
reagent except there are up
to 8 labeled positions
(113–121 Da except 120).
The 305 group composed of
the 185/120 Da pair is not
used because the immonium
ion of phenylalanine occurs
at m/z 120.

FIGURE 3.39 Isotope compositions of the iTRAQ reagents.

ions, released by CID, at m/z 114, 115, 116, and 117, are used to track the peptides from the parallel experiments. The combined isotope compositions of the reporter and balance ions each add the same mass to the peptides; thus, the precursor ion of a peptide selected for CID is the same from all four experimental conditions, while the reporter ion released reflects the relative amounts of the proteins generated during the individual experimental conditions. For absolute quantification, one reagent is reacted with a known concentration of a specific peptide that acts as an internal standard with respect to which the other three sets of proteins can be quantified.

Because all peptides are labeled and only one ion is used to monitor all peptides obtained from each physiological condition evaluated, there is a strong possibility that peptides from different proteins will interfere with each other and yield partially overlapping chromatographic peaks. Therefore, it is important to improve separation, e.g., by using two-dimensional chromatography. The complexity of samples can be reduced by separating cell components, such as organelles and membranes, prior to protein digestion. The low m/z values of the reporter ions render the technique unsuitable for ion traps because these instruments cannot observe progeny ions that are less than one-third of the m/z of the precursor ion (see the "one-third" rule, Section 2.3.2.1).

3.5.1.9.4 Label-Free Quantitative Proteomics

Label-free is a quantification methodology in which an internal standard is used in a manner similar to that used for the quantification of small molecules. Proteins are recovered and digested, and ion chromatograms are extracted (from the total

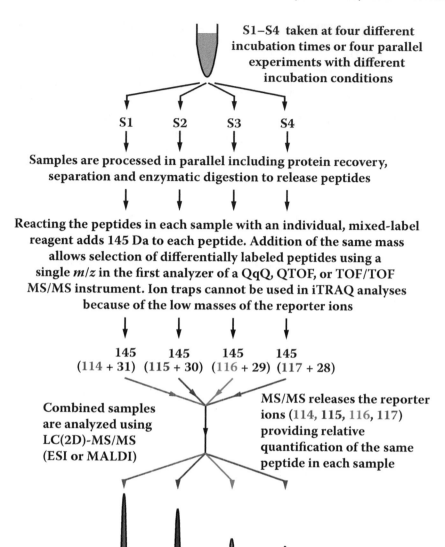

FIGURE 3.40 Quantification using 4-plex iTRAQ (an *in vitro* labeling method).

ion chromatogram) for every peptide produced by LC-MS, followed by the integration of peak areas. Quantification is based on the co-injection of specified amounts of a digested protein of known composition. The identity of the unknown proteins is established from the subsequent MS/MS scans of the peptides. The technique is usable in the 10 fmol to 100 pmol concentration range.

Another quantification approach is based on the fact that the likelihood of observing an MS/MS spectrum of a peptide is a function of its intensity (concentration) and, consequently, of the protein from which the peptide originates. The method uses spectral counting and provides a course measure of concentration with poor incremental sensitivity because only a few peptides are seen for many proteins. Furthermore, compensation has to be made for the molecular mass of the identified proteins because more peptides are generated (when the protein is digested) as the molecular mass of the protein increases. Thus, there will be an increased likelihood of observing peptides corresponding to larger proteins during subsequent MS/MS analysis.

The main advantage of the label-free strategies over methods that use labeling is the possibility of observing the abundances of multiple proteins in a single analysis. These techniques are reliable, versatile, and cost-effective. Difficulties are that rigorous experimental conditions must be maintained, including close control of sample preparation, LC retention times, and injection reproducibility. Another problem is that the ionization efficiencies of ESI are different for individual peptides. Highly reproducible LC is required to determine the proteins from which the peptides originate because of the large numbers of peptides from enzymatic digests may produce co-eluting peptides. Preferred mass spectrometers are those with high resolution and the capability to determine masses with high precision, e.g., TOF, ICRMS, and orbitrap.

3.5.2 NUCLEOMICS (NUCLEIC ACID ANALYSIS)

While there is an amazingly long list of omics subjects, particularly in systems biology, the term *nucleomics* is still not in general use (in 2013). It does seem logical to use this term when referring to all information about the biology of nucleic acids. Information required for DNA ranges from the basic constituents (simple nucleosides and nucleotides) through the chains that delineate individual genes to the extremely long chains that comprise chromosomes (single molecules containing over 200 million base pairs, >12 billion atoms). RNA, the other class of nucleic acids, has multiple functions in protein synthesis (messenger, ribosomal, and transfer RNA). Mass spectrometry has an increasing role in the analysis of a wide variety of oligonucleotides and nucleic acids, including recent promising applications for DNA sequencing, genotyping, and genetic diagnosis.

DNA and RNA are linear biopolymers that form multiply charged envelopes of ions in ESI and singly charged ions in MALDI. When analyzed by MS/MS using CID, both DNA and RNA fragment at the phosphate groups that form the junctions between the nucleosides. Thus, the bases can be sequenced using strategies similar to those in proteomics. The bases are numbered from the 5′ (5-OH) end (B_1, B_2, . . . , B_x) and with respect to the fragmentation that can occur at four sites around the phosphodiester (phosphate) group. The fragments are labeled a, b, c, and d, when starting from the 5′ end, while they are designated w, x, y, and z from the 3′ end, with subscripts corresponding to the distance from the respective end of the molecule, e.g., a_2 refers to a fragmentation after the second base from the 5′ end, while w_3 designates a fragmentation after the third base from the 3′ end (Figure 3.41). The energy imparted to an oligonucleotide during CID may be sufficient to cause fragmentation of the bond between the sugar and the base, resulting in a loss of sequence informa-

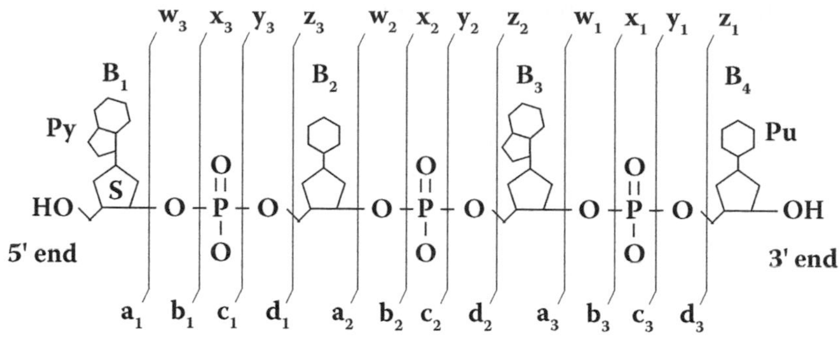

S = sugar: deoxyribose for DNA (ribose for RNA)
PO$_4$ = phosphate group
B = base: numbered from 5' end, attached at 1 position of the sugar
Py = pyrimidine base: adenine and guanine
Pu = purine base: cytosine and thymidine for DNA (uracil in RNA)
Fragments from 5' end labeled a–d
Fragments from 3' end labeled w–z

FIGURE 3.41   Structure and fragmentation points of nucleic acids.

tion. The removal of a base during CID is designated by fragment-B_x (base lost); e.g., a_3-B_2(A) indicates an a_3 fragment from which an adenine has been lost at B_2.

The acidic nature of the phosphate group gives it high affinity for alkali metal ions (Na^+ and K^+). Accordingly, the ions formed in either ESI or MALDI normally include a number of Na and K atoms resulting in spectra that are difficult to interpret, particularly when ESI is used, as there may be several envelopes of multiply charged ions for every nucleic acid strand. Extensive sample cleanup, replacement of the alkali metals with ammonium ions (from which ammonia is released during ionization to yield protonated analytes), and the use of negative ionization can reduce the complexity of the spectra. In negative ion ESI losses of acidic protons from the phosphodiester bonds along the oligonucleotide chains lead to the formation of deprotonated anionic species and the characteristic ESI envelopes of multiple (negative) charge states. The deconvolution techniques utilized in protein analysis can be used for nucleic acid spectra. ESI is effective for the determination of relatively high masses (~30 kDa, ~100 nucleotides).

MALDI is also carried out in negative ionization mode with 3-hydroxypicolinic acid and trihydroxyacetophenone as the favored matrices. Formation of sodium adducts remains a problem with MALDI, often capping the molecular mass that can be analyzed to a chain length of ~100 bases (similar to ESI).

Sequence determination should be simple given that there are only four base types per nucleic acid. Because each nucleotide has a mass of around 310 Da (289–329 for DNA, 305–345 for RNA; amino acid residues average around 120 Da) a segment of a nucleic acid with a given mass will have only about 40% the number of nucleotides as there would be of amino acids in an equal mass of protein. Therefore, less sequence information is obtained from a nucleic acid clip than from a protein (peptide) of the same molecular mass. The masses of the residues of nucleic acid components are given in Table 3.10.

Modifications of DNA usually consist of substituting one base for another, with a consequent change in the molecular mass. Such single-base changes are called *single-nucleotide polymorphisms* (SNPs; pronounced "snips") and occur as often as 1 or 2 per 1,000 nucleotides throughout the genome of the human population, but must occur in >1% of the population to be considered a SNP. The most common substitution is cytosine to thymine. The modifications display a highly stable inheritance. Maps of SNP markers are used in genome-wide linkage and association studies of genetic traits and in pharmacogenomic applications. The smallest alteration found for a SNP (9 Da) corresponds to the change of mass between an adenine (A) and a thymine (T). Attaining the necessary level of mass measurement accuracy is straightforward when using ESI

TABLE 3.10
Residues of Nucleic Acid Components

Bases, Nucleoside, and Nucleotide Residues	Residue Formula	Monoisotopic Mass (Da)	Average Mass (Da)
DNA			
Cytosine	$C_4H_4N_3O$	110.0354	110.0941
Thymine	$C_5H_5N_2O_2$	125.0351	125.1054
Adenine	$C_5H_4N_5$	134.0467	134.1188
Guanine	$C_5H_4N_5O$	150.0416	150.1182
Deoxycytidine	$C_9H_{11}N_3O_3$	209.0800	209.2019
Deoxythymidine	$C_{10}H_{12}N_2O_4$	224.0797	224.2133
Deoxyadenosine	$C_{10}H_{11}N_5O_2$	233.0913	233.2266
Deoxyguanosine	$C_{10}H_{11}N_5O_3$	248.0862	249.2260
Deoxycytidine-monophosphate	$C_9H_{12}N_3O_6P$	289.0464	289.1818
Deoxythymidine-monophosphate	$C_{10}H_{13}N_2O_7P$	304.0460	304.1932
Deoxyadenosine-monophosphate	$C_{10}H_{12}N_5O_5P$	313.0576	313.2065
Deoxyguanosine-monophosphate	$C_{10}H_{12}N_5O_6P$	329.0525	329.2059
RNA			
Cytosine	$C_4H_4N_3O$	110.0354	110.0941
Uracil	$C_4H_3N_2O_2$	111.0194	111.0788
Adenine	$C_5H_4N_5$	134.0467	134.1188
Guanine	$C_5H_4N_5O$	150.0416	150.1182
Cytidine	$C_9H_{11}N_3O_4$	225.0750	225.2013
Uridine	$C_9H_{10}N_2O_5$	227.0668	227.1940
Adenosine	$C_{10}H_{11}N_5O_3$	249.0862	249.2260
Guanosine	$C_{10}H_{11}N_5O_4$	281.0760	281.2248
Cytidine-monophosphate	$C_9H_{12}N_3O_7P$	305.0413	305.1812
Uridine-monophosphate	$C_9H_{11}N_2O_8P$	306.0253	306.1660
Adenosine-monophosphate	$C_{10}H_{12}N_5O_6P$	329.0525	329.2059
Guanosine-monophosphate	$C_{10}H_{12}N_5O_7P$	345.0474	345.2053

on TOF or FT instruments as 1 Da ≈ 30 ppm for a nucleotide with 100 bases; this degree of accuracy is not available with MALDI-TOF systems.

The ability to determine molecular masses accurately permits the rapid assessment (without full sequencing) of polymorphism for DNA strands of defined length from different origins (individuals). Once a change in molecular mass is detected, the localization of an SNP within that nucleic acid fragment requires the determination of the nucleotide sequence. Alterations of the sequence without a change in molecular mass can arise when two SNPs counteract each other in the sequence. The frequency of SNPs and the lengths of nucleic acids that can be analyzed by mass spectrometry make it improbable that two counteracting SNPs will occur and therefore be missed.

Peptide nucleic acids (PNAs) are a class of synthetic DNA mimics where the sugar-phosphate backbone has been replaced by repeating N-(2-aminoethyl)glycine units linked by peptide bonds, and the bases are present in specifically designed sequences. These compounds bind in PNA:DNA pairings according to their sequence, with bond strengths greater than DNA:DNA binding, making them useful molecular probes. Both MALDI and ESI effectively ionize these compounds, forming positive ions. Structural information (including sequencing) is obtained using MS/MS.

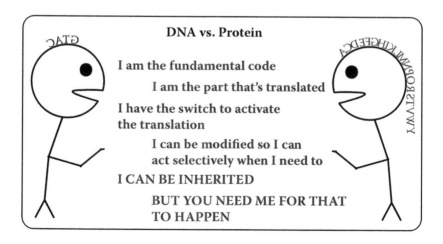

3.5.3 Glycomics

Carbohydrates (*saccharides*) consist of only carbon, hydrogen, and oxygen, and can be classified into three groups: simple *mono-* and *disaccharides* (sugars), those with branched structures (*glycans*), and long-chain *polysaccharides*. The last are free species and include compounds such as hyaluronic acid and heparin. A *glycome* is the total complement of sugars of an organism, whether present freely or as part of complex molecules. Glycobiology encompasses the study of all the carbohydrates. *Glycomics* refers to the study of the glycans in a cell, tissue, or organism.

The glycans consist of monosaccharides linked by O-glycosidic bonds (residue ions for biologically important monosaccharides are listed in Table 3.11). They may be homo- or heteropolymers with linear or branched structures (Figure 3.42). Those

TABLE 3.11

Residues of Biologically Important Monosaccharides

Monsaccharide Residues	Residue Formula	Monoisotopic Mass (Da)	Average Mass (Da)
Pentoses	$C_5H_8O_4$	132.0423	132.1146
Deoxyhexoses	$C_6H_{10}O_4$	146.0579	146.1412
Hexosamines	$C_6H_{11}NO_4$	161.0688	161.1558
Hexoses	$C_6H_{10}O_5$	162.0528	162.1406
Glucuronic acid	$C_6H_8O_6$	176.0321	176.1241
N-Acetylhexosamines	$C_8H_{13}NO_5$	203.0794	203.1925
N-Acetylneuraminic acid	$C_{11}H_{17}NO_8$	291.0954	291.2546
N-Glycolylneuraminic acid	$C_{11}H_{17}NO_9$	307.0903	307.2540

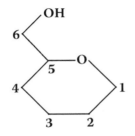

Numbering of the carbon positions for a hexose.

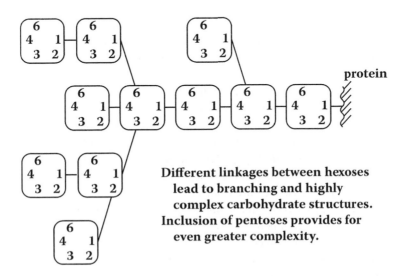

Different linkages between hexoses lead to branching and highly complex carbohydrate structures. Inclusion of pentoses provides for even greater complexity.

FIGURE 3.42   Structure of oligosaccharides.

with branched structures (*oligosaccharides*) are usually associated with proteins forming *glycoproteins*. This family of proteins often occurs as components of the intercellular environment, i.e., at the interfaces between cells, and also at the point of interaction between cells and externally derived molecules such as nutrients and toxins. Glycoproteins are of major importance in cellular communication and in disease states.

Glycosylation is a form of post-translational modification in which an oligomeric sugar (often 7 to 25 monosaccharide units) is added to a specific site of a protein in a process controlled by enzymes such as glycosyltransferases and glycosidases. It is estimated that about 50% of proteins in mammalian cells are glycosylated at any given time. Proteins may exist in various glycoforms. The presence of different glycans on the surface significantly impacts the properties of the proteins including their charge, conformation, and stability. The longer glycan chains of glycoproteins may protrude far from membranes, while the shorter glycan chains of *glycolipids* (<12 monosaccharide units) usually remain in the immediate vicinity of the cell surface. Abnormal glycosylation has been correlated with an unusually wide variety of diseases, ranging from congenital abnormalities to cancer.

There are several ways that glycans can be coupled with a protein. N-linked glycans are formed via the nitrogen of the asparagine or arginine side chains. O-linked glycans are attached through the hydroxy oxygen on the side chains of serine, threonine, tyrosine, hydroxylysine, or hydroxyproline residues of proteins, or via the lipid oxygens in glycosylated lipids. Phosphoglycans are formed when the carbohydrate is linked to the phosphate of phosphoserine.

There is a significant difference between glycosylation, as described above, and *glycation*, where a sugar molecule is added to a protein or lipid via a reductive (nonenzymatic) reaction that results in the nonspecific addition of monosaccharides to proteins. Glycation has significant health effects because it is a nonreversible process that leads to the *aging of tissues* through reduction of flexibility.

The fact that the oligosaccharides have three-dimensional, tree-like structures is a particular challenge to their analysis by mass spectrometry because of the need to determine the points of connectivity in each chain, the location of branch points, and the sequence of the sugar residues. The number of structural permutations is several orders of magnitude higher for the three-dimensional composition of oligosaccharides than for peptides where the amino acids are assembled linearly. For example, 20 different monosaccharides could form ~10 million linear or branched trisaccharides, in contrast to the ~6,800 tripeptides that can be assembled from the 20 amino acids. An additional problem is that many sugars are difficult to fully characterize by mass spectrometry because they are isomeric.

The nomenclature for CID fragmentation of oligosaccharides is shown in Figure 3.43. Fragmentation can occur on both sides of the glycosidic bond, Y (B) and Z (C), and across the ring, X (A). Subscripted numbering of the X, Y, and Z ions is from the linkage to the peptide, starting at 0, while for the A, B, and C ions it is from the terminal sugar starting at 1. Superscripted numbering designates the bonds that are broken in cross-ring fragmentation, with zero being the first bond clockwise from the oxygen.

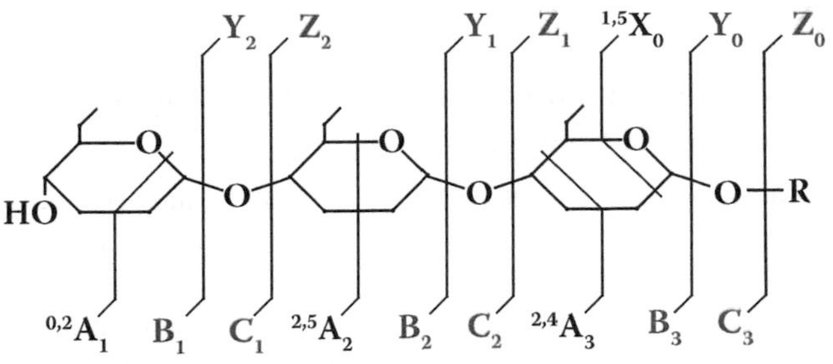

Primary fragmentation is at the glycosidic bonds,
Y (B) and Z (C).
Labeling is from the reducing end.
Cross ring fragmentation, X (A), is labeled by bond number
clockwise from the ring oxygen with the first bond being 0.

FIGURE 3.43 Fragmentation of carbohydrates.

Mass spectrometric analysis of glycoproteins has three objectives: identify the protein substrate, determine the site of glycosylation, and characterize the structure (i.e., composition, sequence, and linkage) of the glycans.

The complexity of glycans requires the utilization of all available tools in glycoprotein analysis. Initial separations can be made using LC, CE, or anion exchange chromatography. CE is particularly useful for separating isomers. ESI and MALDI are both used, in combination with low- and high-energy CID, MS/MS, and MS^n. Accurate mass measurement is valuable because the high oxygen content of sugars suppresses the mass sufficiency (mass above the unit mass value) of glycopeptides, thus providing an indication that an unknown peptide may be glycosylated.

The determination of the branched linkages of the oligosaccharides requires multiple mass spectrometric techniques. Positive ionization is used to generate protonated species if the amino sugar forms a neutral or basic glycoconjugate. Negatively charged deprotonated ions are utilized for sulfate-containing acidic glycoconjugates. The efficiency of ionization is improved significantly by permethylating the sugars to increase their hydrophobicity, volatility, and surface activity. Furthermore, doping the samples with certain metals can stabilize different parts of the sequences, particularly at the glycosidic bonds, leading to alternative fragmentation pathways. All alkali metals (lithium to cesium), as well as divalent species such as calcium and magnesium, can be used for doping. Peracetylation can direct fragmentation toward specific bonds, and high-energy CID can promote cross-ring fragmentation. Examples of different experimental conditions to obtain structural information for oligosaccharides are shown in Table 3.12. Fragmentation of the secondary (CID) ions to obtain MS^3 spectra can provide additional structural data when ion traps are used.

Selective enzymatic digestion of the trypsin-derived peptides is another approach for obtaining structural information from glycoproteins. Sequential enzymatic removal of individual nonreducing terminal sugar residues is followed, after each digestion, by mass spectrometric analysis. These strategies provide data regarding

TABLE 3.12

Examples of Structural Information for Oligosaccharides Provided by Different Experimental Conditions

Derivatization/Metal Presence	Ion Polarity	MS/MS	Information Generated
None	Positive/negative	CID	B and Y fragmentation for sequence
None	Negative	HCID	Cross-ring fragmentation (A/X)
Sodium	Positive	HCID	Cross-ring fragmentation (A/X)
Lithium	Negative	CID	Differentiation of isomers
Alkali metals: Li to Cs	Positive	CID	Choice of metal controls extent of glycosidic fragmentation
Earth metals: Ca and Mg	Positive	CID	Mg favors cross-ring (A/X) fragmentation
Per alkylated: Methylation Acetylation	Positive	CID	Distinguish branched and linear species
Permethylation	Positive	CID	Determine sugar linkages

Note: CID = collision-induced dissociation, HCID = high-energy CID.

the heterogeneity of the oligosaccharides at each site, the identity of the removed sugar, and the composition of the carbohydrate remaining on each glycopeptide.

3.6   IMAGING MASS SPECTROMETRY

The preparation of samples for mass spectrometric analysis almost always requires recovery of the analyte from a matrix, often followed by further preparative steps prior to introduction of the sample into the ion source. Such destructive processes are unacceptable when the objective is to determine the specific location of a particular compound within a biological sample, e.g., the location of a certain lipid in a tissue, and led to the development of *imaging mass spectrometry* with MALDI as the ionization method, although DESI is also applicable.

Histology is a visual process, assisted by differential staining, used to characterize biological tissues and often aimed at diagnosing a disease, e.g., to differentiate the border of a tumor from disease-free tissue. However, histology provides virtually no information about the molecular composition of the tissue (or tumor) involved. The ability to differentiate between healthy and diseased tissue would improve markedly if molecular markers of diseases could be located and identified. Imaging mass spectrometry provides *molecular histology* that complements conventional *visual histology*. Most compound classes can be investigated, including proteins, peptides, and lipids, as well as drugs and their metabolites. Initial efforts have concentrated on lipids, given that these compounds are major cellular components that can be ionized readily from surfaces.

Imaging is usually carried out on frozen tissue sections (~20 microns thick) placed on a MALDI plate. The chemical nature of the compounds of interest determines the next steps in sample preparation, including the choice of the MALDI matrix (alpha-cyano for small molecules and sinapinic acid for proteins) that is sprayed or sublimed onto the tissue surface. Any additional sample preparation, e.g., treatment with enzymes to release peptides for subsequent MS/MS analysis and protein identification, should be completed prior to deposition of the matrix.

The acquisition of mass spectra from a tissue section can take several hours, as the laser spot raster is built up over the tissue. Sophisticated data processing is required to evaluate the large data files generated, aiming to determine the changes in *m/z* values across the tissue and establish whether relative changes in the intensities of specific *m/z* occur between different points within selected sections, e.g., between diseased and healthy areas (Figure 3.44). Data evaluation strategies use sophisticated statistical approaches, such as principal component analysis and hierarchical differentiation.

The *spatial resolution* obtainable within tissues, that are now approaching the cellular level, is a compromise between the size of the laser spot and the sensitivity of the mass spectrometer (Section 4.13). The mass range over which MALDI can be used is limited by interferences from the matrices; the lower limit is 500–600 Da. DESI is not limited at lower masses and can facilitate analysis of compounds with masses of <500 Da.

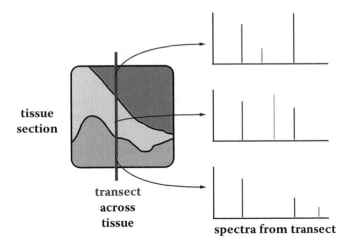

Spectra taken on a transect across a tissue section vary.
   Some compounds are unique to certain parts of the tissue and
   may be diagnostic.
   Some compounds are always present but vary in concentration.
   Some compounds are present always at the same concentration.
An image of the tissue is reconstructed from the spectra of
   multiple transects obtained by rastering the tissue.
Using MALDI, each peak in the spectra corresponds to a different
   compound.
Sophisticated data processing is required to identify biomarkers.

FIGURE 3.44   MALDI imaging of an intact tissue section.

3.7 BIOINFORMATICS

The rather vague term *bioinformatics* is a component of the new field of *information science* that uses statistical and other mathematical techniques to provide interpretations of experimental results obtained in the study of various problems. In bioinformatics the emphasis is on determining correlations in biology and medicine, e.g., in the prediction of disease probabilities. Bioinformatics is rapidly becoming a scientific disciple of its own, and it is not an easy one.

The need for new approaches/techniques for data interpretation arose because modern mass spectrometric techniques generate awe-inspiring amounts of data. The days when a single mass spectrum provided an answer to a specific problem are over, and spectra, even of a single compound, are often created by multiple ionization methods and may be so different from each other that it can be difficult to accept that they have originated from the same analyte. Thousands of individual mass spectra are generated (and recorded sequentially) even when using a single technique such as GC-MS. When it comes to complex analyses such as LC-MS/MS, with continuous acquisition of both MS and MS/MS spectra, including accurate mass measurement and even ion mobility data, individual data files can be 5–10 gigabytes and, consequently, mass spectrometry laboratories can generate terabytes of data monthly.

In mass spectrometry, bioinformatics is being used increasingly in virtually all applications described in this chapter (and also in Chapter 4), particularly in proteomics (sequencing, structure, function, and interaction), metabolomics, and nucleomics. There is an increasing need for mass spectrometrists to become familiar with, or at least to understand, what is happening to the data they obtain when analyzing environmental and biological samples. Although a detailed discussion is way beyond the scope of this book, a brief description of the basic concepts and methods is relevant.

The generic term *data mining* refers to computer-based searching for recognizable patterns within huge data sets. This is the first step in transforming bits into knowledge. Common processes include *data conversion*, *feature detection*, *alignment* (elimination of retention or migration shifts between data sets), *normalization* (elimination of systematic bias), *identification* (matching candidate spectra with standards), and *quality control* (using benchmark tests).

The first step in data mining, for instance, of the results in a metabolomics study, is *multivariate analysis*, where the aim is to reduce the volume of data into a few dimensions for subsequent classification and outcome prediction. It is noted that identification of compounds is not necessary, only *m/z* and intensity data is needed for this type of discrimination analysis. No prior knowledge of anything (i.e., of the data) is required to build *unsupervised* statistical models. *Principal component analysis* (PCA) is the most commonly used unsupervised technique. In contrast, separate *training sets* of data must be obtained to build *supervised* learning models used for prediction or classification. Techniques involving *data splitting* (e.g., cross-validation) have been developed for situations where independent data sets are not available for training and validation. The most commonly used supervised learning techniques include *partial least squares discriminant analysis, support vector machines, random forest methods, artificial neural networks,* and *classification and regression trees.*

Protein research. Data processing is a critical component of protein identification (and quantification). A plethora of informatics tools have been targeted toward database searches, sequence comparisons, and structural processing of MS-based proteomic data. The most common approach is based on *database searching,* where there is comparison of experimentally determined peptide masses and MS/MS data to libraries. The latter are composed of previously submitted results and theoretical peptide masses derived from genomic data. The most common, a hybrid approach, to protein identification uses short (three to five residues long) *sequence tags* (combined with peptide masses) extracted from the spectra and then used in error-tolerant database searching. The list of candidate proteins is next ordered using *scoring algorithms* with various levels of sophistication. In *de novo* sequencing, peptide sequences are read out directly from fragment ion spectra. Instrument manufacturers have created sophisticated software packages to undertake protein identification. The field has expanded to the point that there are independent software developers who have created programs that can utilize the raw data from multiple instrument platforms. In addition, there are lists of publicly available software programs for MS/MS-based proteomics, including tools for database searches, spectral matching, *de novo* sequencing, statistical validations, database storage, mining and sharing, and quantification.

There has been a consolidation of the protein information into the UniProt and NCBI databases, the two primary sources of protein sequence data. The UniProt (Universal Protein Resource) database, more properly called the UniProtKB (UniProt Knowledgebase), includes the Swiss-Prot and TrEMBL data. This database (based in the United Kingdom) is operated by a consortium of the European Bioinformatics Institute, the Swiss Institute of Bioinformatics, and the Protein Information Resource. The Swiss-Prot section of the database is subject to manual curation to improve the quality of the record, and provides other available relevant information about some proteins. Data are included for species of origin, function, subcellular location, structural variants, etc. The increasing rate of data generation has overwhelmed the capacity for manual curation and led to the formation of the TrEMBL database that contains many more entries than Swiss-Prot, but is computer annotated and has not yet gone through a more rigorous manual evaluation. The NCBI database is operated by the National Center for Biotechnology Information, a component of the U.S. National Institutes of Health. The information in the NCBI database is similar to that in the UniProt system but has additional features, including crystal structures and genetic data.

A new subdiscipline of bioinformatics, *comparative proteogenomics,* is based on the possibility of mapping sets of sequenced peptides onto the positions in the genome that code for them, i.e., integration of data from MS-based proteomics with DNA sequence data sets. Databases and software to determine if the available data are over- or underrepresented include *Gene Ontology, protein domains,* and *pathway databases.* Another application (of many) is the combination of MS-based proteomic data with information from *transcriptomics* to study the processes and regulation of cellular functions.

The importance of the full identification of putative protein markers is controversial in biomarker studies. It has been suggested that the identity of the markers need not be known as long as consistent protein profiles can be obtained that represent differential expression(s). In other words, diagnosis should be considered a prediction without being concerned with etiology. When there are no individual peaks or groups of peaks with intensities significantly different between normal vs. pathologic samples, the bioinformatics algorithm needed for diagnosis must be based on a supervised approach based on training data sets.

3.8   BUYING A MASS SPECTROMETER

"Continual cravings for new mass spectrometers are a genetic predisposition for our species."

(Kenneth I. Busch)

Mass spectrometry is big business. Sales were about $3.9 billion in 2011 (about 30% of the total spectroscopy market) and may rise to $4.8 billion by 2014. Growth in 2014 is estimated at 7.9%. The most common purchases in 2011 were tandem LC-MS (33%), GC-MS (20%), and LC-TOF-MS (16%). Rapid increases have been noted in high-resolution MS, particularly FT-MS (because of more accessible technologies, e.g., orbitrap), as well as in ambient ionization technologies, e.g., DART. Some 80% of the market is shared by AB Sciex (Framingham, Massachusetts), the market leader; Thermo Fisher Scientific (Waltham, Massachusetts); Agilent Technologies (Santa Clara, California); Waters (Milford, Massachusetts); and Bruker (Billerica, Massachusetts), each manufacturing, marketing, and supporting a wide variety of MS techniques. Three other companies, Shimadzu (Columbia, Maryland), Leco (St. Joseph, Michigan), and Jeol (Peabody, Massachusetts), are also well known, particularly for specific applications, notably MALDI-TOF, GC/GC-TOFMS, and magnetic sector instruments.

The process of buying a mass spectrometer consists of a number of consecutive steps: (1) define the objectives in both broad and specific terms; (2) determine which instrument and data system would be ideal/adequate to attain the objectives (use the literature and contact experts); (3) establish the maximum budget available; (4) contact several MS manufacturers and ask for detailed proposals, including prices, performance values (general as well as with respect to the stated objectives), warranties, delivery, and other pertinent information, including contact information for other users of similar instruments; and (5) arrange for appropriate demonstration in the manufacturer's applications laboratories.

The first, and probably the simplest, situation is when the purchase is protocol driven for a specific, but routine, project that is large enough to demand and justify a dedicated instrument, e.g., for an assay that is part of a regulated work flow. The choice of mass spectrometer is straightforward when there is an already published and accepted technique for the analyte(s) involved. Often, however, the project demands modified or new methodologies; thus, the instrument will be used for research, albeit with a relatively simple objective. Maximum achievable sensitivity is, of course, invariably the primary requirement.

The second scenario is the acquisition of a mass spectrometer for general research purposes, including existing, contemplated, and dreamed about projects, for support of a wide variety of research activities in an academic department or industrial organization, often including pharmacological, biological, medical, or environmental research. This means that the use of the mass spectrometer will be "problem driven," and the mass spectrometer is expected (mostly by administrators) to be able to do "everything."

The third situation is when the objectives (usually specific and limited in scope) require the highest resolution (>1,000,000) that is available only with FT-ICRMS systems. Because the timescales of data acquisition for these instruments are not compatible with LC, the current approach is to bypass the chromatographic step and use flow injection to introduce samples, and then use an LIT (often included) in conjunction with the resolving power of the FT-ICR to "tease apart" highly complex mixtures (e.g., top-down proteomics or crude oils). It is likely that the organization contemplating such an instrument already has other mass spectrometers along with experienced operators/researchers.

The choice of the *sample introduction* system is determined by the polarity of the analyte(s) and the complexity of the matrix in which the sample is present. Nonpolar compounds are suitable for GC. Examples include environmental pollutants (e.g., PCB), hydrocarbons, and compounds that have been derivatized to mask polar groups (e.g., fatty acids). For limited mixtures and pure nonpolar compounds, a solids probe or heated gas inlet can also be used for sample introduction. Much more frequently, the analytes are polar compounds requiring LC-based sample introduction (or flow injection for limited mixtures or pure compounds). Examples include natural products, endogenous compounds, and drugs. A further class is the biopolymers (proteins, carbohydrates, nucleic acids) that comprise a very large, and increasingly important, group of polar compounds where high sensitivity is usually a major requisite. Sample introduction is via nano-LC, the choice of which is based on the often forgotten fact that MS is a *concentration-dependent* technique.

The type of *ionization method* is determined by the method of sample introduction. The most useful sources with GC are EI and CI, and with LC, ESI (or another API method, e.g., APCI or APPI). MALDI is a stand-alone ionization method that is best suited to peptides and proteins. Because MALDI is an off-line technique, samples can be investigated multiple times as opposed to LC, where peaks cannot be revisited without reinjecting the sample. Although it is highly desirable to have multiple ionization sources, the budgets for dedicated instruments may not allow such luxuries.

The choice of the *mass analyzer* (the most expensive component of a mass spectrometer) is relatively straightforward for dedicated applications. Quadrupole analyzers, in SIM mode, and triple quadrupoles (QqQ), in SRM mode (MS/MS), are the best for sensitive quantification of small molecules (with high specificity). With single-analyzer instruments the amount injected is in the picogram to nanogram region, while the QqQ reaches into the femtogram or lower range. The QqQ also provides product, precursor, and neutral loss scans (MS/MS), enabling the search for families of compounds such as metabolites. However, with this versatility comes a loss of sensitivity because quadrupoles are scanning instruments and, therefore, have a poor duty cycle compared to nonscanning analyzers, e.g., TOF.

For general purpose instruments (scenario 2), the most important parameters are resolution and accurate mass measurement for the identification and characterization of unknowns. Typical are MS/MS systems based on TOF or FT (orbitrap or ICR). The major categories here are TOF/TOF, QTOF, LT-TOF, LT-orbitrap, and LT-FT-ICR. For analytes of >1 kDa, TOF/TOF systems in combination with MALDI sources are straightforward and should be evaluated where multiple users are involved. QTOF, LT-TOF, and LT-orbitrap are all suited to LC operation at regular or nanoflow rates. The resolution of orbitraps is much higher (250,000) than that of TOF instruments (20,000 to 60,000), although available resolution on the former is subject to duty cycle limitations. Table 3.13 lists representative examples of the types of instruments one might consider for common applications.

A major consideration in all instrument purchases is the *post-processing* of data, ranging from the simplest cases, where the objective is quantification of known compounds, through accurate mass determinations at the trace level, to complex data files that are in the gigabyte range. Instrument manufacturers often offer large and sophisticated (and expensive) data packages, and these should be considered and thoroughly evaluated. Separate computers (64-bit systems with 16 GB or more memory) with sophisticated proprietary software are often needed for interpretative studies and for the use of databases on the Internet.

Budgeting, a major consideration, is a three-component how-to process: find funds, convince management, and bargain with the vendors. For the first two, the best advice is to become a scholar of Machiavelli. Bargaining with the vendors also has three components: determining the total (final) purchase price, uncovering how much equipment/software can be squeezed into the agreed price, and establishing what is and is not included in the warranty period as well as in any subsequent maintenance contract. Establishing a rapport with the representatives and

TABLE 3.13

Examples of Types of Instruments to Consider for Common Applications of Mass Spectrometry

Objective	Type of Mass Spectrometer (examples given, not exhaustive)
Quantification of drugs/metabolites in plasma	LC-QqQ
Quantification of persistent environmental pollutants	GC-Q, GC-QqQ
Protein identification	nanoLC-QTOF, nanoLC-LT-orbitrap, nanoLC-spotting robot-MALDI-TOF/TOF
Top-down analysis of large proteins	FT-ICRMS
Imaging technique of tissues	MALDI-TOF/TOF
Petroleomics, humates, fulvates	FT-ICRMS
Surface analysis	DESI or DART (compatible with multiple analyzers)

their supervisors is very important because there is always some leeway on prices as well as on promises about the degree and length of user training. Representative prices (United States) for major instrument types are given in Table 3.14. Budgeting should also include the cost of yearly maintenance (about 10% of purchase price for a service contract) and the inevitable cost of parts. There are also the so-called running expenses, ranging from vacuum pump oils and gases to computer and printer supplies. Yet another expense is for data systems and software from independent vendors. FT-ICR systems are a special case because they require constant supplies of both liquid helium, a commodity that is becoming increasingly expensive because of supply limitations, and liquid nitrogen to maintain the superconductive magnets.

Still another important component of the decision process is the evaluation of buyer-submitted samples by the vendors, particularly when more than one manufacturer is considered. The samples to be submitted for analysis should be designed so that the results will confirm the stated performance/capabilities touted by the manufacturer. To avoid disappointments, the samples should not be unknowns, and one should not expect the results to meaningfully contribute to any research project. The application (demonstration) laboratory of the manufacturer is likely to allot only one or two days to analyze your samples. A visit to the manufacturer's site should be arranged to coincide with the time when the test samples are analyzed. Such

TABLE 3.14

Representative Prices (in 2013) for Major Instrument Types

Instrument Type	Cost (US$)
Q	80 K
TOF, ion trap, QqQ	150–350 K
TOF/TOF, QTOF, orbitrap	400–900 K
FT-ICR	>1 M

trips provide an important opportunity to talk to, and establish personal relationships with, individuals with a variety of knowledge/functions within the company, particularly when the purchase of a high-end instrument is contemplated.

A relatively easy way to determine the truly essential components and capabilities and drawbacks of a selected mass spectrometer is to consult current users. Touchy questions may arise when, despite all negotiations, the available budget is just not adequate for a desired instrument. The chosen instrument should have all required capabilities. It is usually not prudent to hope for later upgrading.

Buying a *used/refurbished* mass spectrometer has only one real advantage, and this may be a big one, a cost savings of 30–80% compared to purchasing a new instrument with comparable specifications. Dealing directly with the previous owner is probably not a good idea regardless of the financial benefit. A better approach is to do business (very carefully) with established resellers. Probably the best scenario is to contact mass spectrometer manufacturers who often have refurbished trade-ins or new instruments still in stock at the time new models are introduced. A brief list of items to be concerned with, beyond the obvious questions and requested verifications pertaining to performance (sensitivity, resolution, mass range), includes status and age of vacuum pumps, chromatographs, and interfaces, available ion sources, and accessibility of spare parts/components. Deal-making items (beyond price) should include the type of warranty (3 months is inadequate) and cost of delivery, installation, and training. With respect to these items, established instrument manufacturers almost always offer better deals than direct sellers or resellers/brokers.

A new mass spectrometer should not be considered a magical solution to complex research projects. The decision process may take several months, and there are likely to be ups and downs. Still, purchasing a new (or yet another) mass spectrometer should always be an exciting, even joyful occasion.

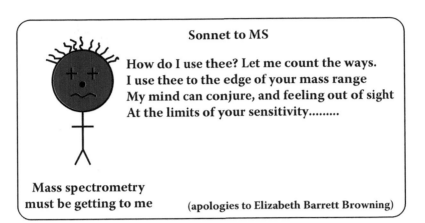

Sonnet to MS

How do I use thee? Let me count the ways.
I use thee to the edge of your mass range
My mind can conjure, and feeling out of sight
At the limits of your sensitivity.........

**Mass spectrometry
must be getting to me**

(apologies to Elizabeth Barrett Browning)

4 Examples from Representative Publications

The first three chapters of this book provided a general overview of mass spectrometry, including information on the structure and function of the components of instruments and how they are combined into instrument systems, as well as of the analytical techniques/strategies available and the types of data generated. This chapter includes brief descriptions of 13 published papers that (although they may appear to be arbitrarily selected) were chosen to provide examples of the diversity (breadth) of mass spectrometry and the wide variety of applications. Some of the papers are straightforward; others are much more complicated. Readers are encouraged to read the original articles.

4.1 AN OPEN-ACCESS MASS SPECTROMETRY FACILITY

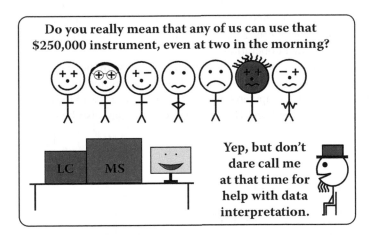

Field: Wide variety of applications, e.g., synthetic chemistry.

Technique: Open-access instrumentation, electrospray ionization–mass spectrometry (ESI-MS), gas chromatography–mass spectrometry (GC-MS), matrix-assisted laser desorption/ionization (MALDI).

Reference: J. Greaves, Operation of an academic open-access mass spectrometry facility with particular reference to the analysis of synthetic compounds, *J. Mass Spectrom.* 37:777–785 (2002).

Mass spectrometers are often used to produce routine data that are still essential components of research programs. Because such data are often integral components of a sequence of experiments, the results are usually required immediately, and delays can become rate limiting.

Mass spectrometers were often the last portion of an analytical sequence that included thin-layer chromatography (TLC), infrared (IR), UV, and nuclear magnetic resonance (NMR). In synthetic chemistry MS was frequently used only to obtain accurate mass values because journals required such data. Open access allows mass spectrometry to be moved to the front of the analytical chain with rapid determination of the molecular masses of analytes.

When magnetic sector instruments dominated mass spectrometry and samples were introduced via solid probes using vacuum locks, there was a need for experienced instrument operators. The decoupling of the mass spectrometer from the dedicated operator began with the development of GC-MS, particularly with the introduction of fused silica GC columns that could be connected directly to the ion source (and were physically robust). However, it was the advent of the atmospheric pressure ionization (API) methods that opened mass spectrometry to a wide audience by enabling the analysis of a much broader range of compounds, particularly polar species, including peptides and proteins (and no vacuum locks were required). Instruments also became more reliable, with most aspects of their operation controlled by computers, including automated tuning, calibration, sequencing of events during analyses, and data acquisition.

It is argued in this paper that routine, rapid generation of data in an open-access mass spectrometry laboratory can make the data acquired timely and thus increase its relevance. To achieve this goal, it was important to simplify both experimental protocols and instrument operation.

Open-access software provides inexperienced users with the opportunity to rapidly analyze samples themselves on a 24/7 basis. It is often convenient to put the open-access operation on a separate computer(s) where there is also access to the data files that can then be examined by the users (Figure 4.1). Reports may also be emailed, although

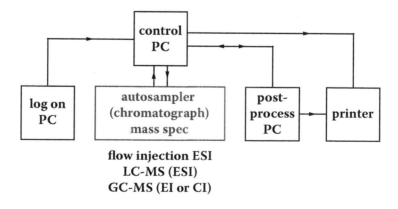

FIGURE 4.1   Layout of an open-access instrument.

it is argued that having to collect reports in the facility increases the opportunity for novices to have useful interaction with experienced instrument operators.

For synthetic chemists, who used to look at mass spectrometry as a technique of last resort because of the delays in obtaining data, open access means that they can now obtain data rapidly. In fact, users come to rely on having access to the instrument at all times. The author had initially planned to allocate certain days for flow injection analysis while other days would be for LC-MS. It soon became clear that flow injection was more popular, and that immediate access to results and rapid sample turnaround was more important than component separation. Access to the technology also altered the strategies some chemists used in their syntheses, e.g., enabling them to undertake reactions on a submilligram scale. A major benefit of the open access strategy is the immediate determination of whether or not the expected molecular mass has been obtained, even though the actual arrangement of atoms within the molecule (connectivity) cannot be determined. The small amounts of sample needed (e.g., a TLC spot can be scraped off a plate, extracted, and analyzed) also means that the progress of individual steps in a reaction scheme can be monitored, rather than having to stop the reactions at arbitrary points only to find that they have not gone to completion.

It is postulated that ESI is preferable to atmospheric pressure chemical ionization (APCI) because of the reduced fragmentation and extended mass range of the former. Also, when ESI is used in flow injection analysis (FIA), the strategy can be simplified by using methanol, without water or acid, as the solvent. FIA (with methanol) does tend to result in the formation of $[M + alkali\ metal]^+$ ions (the alkali metal is usually sodium) rather than $[M + H]^+$ ions, but users soon become familiar with such information. However, it is important to remember that data may be distorted because ESI favors polar compounds; e.g., for a mixture containing 90% of a nonpolar compound and 10% of a polar compound, the polar species will apparently be the major component.

When the analytes require separation, one should use generic chromatographic columns, such as C_{18} in liquid chromatography–mass spectrometry (LC-MS) and 95% methyl:5% phenyl siloxane in GC-MS. The chromatographic methods should be designed to remove as many components as possible from the column to avoid "donating" them to the next user. Thus, in reversed-phase LC-MS the elution gradient should always culminate at high organic composition (e.g., 95% acetonitrile or methanol) for sufficient time to ensure that everything has eluted from the column. In GC-MS, column cleaning is achieved by increasing the temperature for every sample to an appropriate limit, such as 300 °C. While these requirements do slow the sample turnaround time; they are necessary for the integrity of the data.

Many modern instruments, e.g., time-of-flight (TOF) systems, can be used to determine accurate masses; however, it is often questionable whether such numbers are important enough to justify the necessarily stringent calibration, lock mass, and concentration requirements, as well as the more advanced training of users. This is particularly true when MS is moved to the beginning of the analytical sequence, when knowledge of the nominal mass is often sufficient. For instance, accurate mass measurement is often irrelevant when the objective is to synthesize a compound of molecular mass 359 Da but the spectrum obtained shows that the molecular mass of the reaction product is 321 Da.

While the simplest approach may not always be the ideal one, open access makes the case for "pleasing most of the people most of the time." The mass spectrometry facility described has 200–300 users who analyze over 20,000 samples annually. For most of these scientists, this is their first personal access to MS. Once the "I need a molecular mass now" population has been accommodated, more sophisticated questions can be undertaken in conjunction with the staff of the facility. Other open-access analyses may include: (1) quantification, for which there are LC-QqQ-MS systems with appropriate open-access software (one has been added to the facility since the paper was written) and GC-MS systems; (2) identification of unknowns using LC-TOF-MS and GC-MS; and (3) the possibility of analyzing air-sensitive samples.

MALDI instruments, particularly those with automated plate loading, can also be operated successfully as open-access systems. Recently, the MALDI-TOF system in the facility was replaced with a MALDI-TOF/TOF instrument so that interested researchers could undertake peptide sequencing (and other structural determinations) along with standard MALDI-TOF analyses.

A further advantage of open-access facilities is that students and researchers learn about the remarkable diversity and utility of mass spectrometry, knowledge they can carry forward as their careers progress.

4.2   ENVIRONMENTAL: ORGANOCHLORINES IN FISH

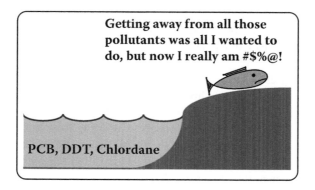

Field: Environmental analysis.

Techniques: GC-MS, negative ion chemical ionization, selected ion monitoring.

Reference: D. L. Swackhamer, M. J. Charles, and R. A. Hites, Quantitation of toxaphene in environmental samples using negative ion chemical ionization mass spectrometry, *Anal. Chem.* 59:913–917 (1987).

Background: Toxaphene is a complex mixture (>670 compounds) of chlorinated bornanes and bornenes that was used as an insecticide for ~40 years from 1945. Many other chlorinated compounds entering the environment over the same period

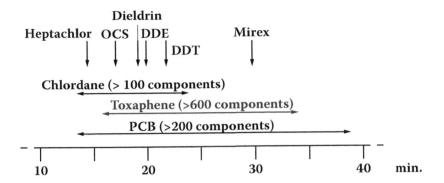

FIGURE 4.2   GC retention times of organochlorines.

were complex mixtures, such as chlordane (>100 compounds) and PCB (209 congeners), while other contaminants were single compounds, such as heptachlor, octachlorostyrene, dieldrin, and DDT. Any or all of these compounds can occur in the extracts of environmental samples; thus, some 1,000 chlorinated compounds may be present, way beyond the number (~100) that can be separated by GC. The overlapping retention times of these compounds are illustrated in Figure 4.2.

Methods: In the method reported, mass provides a second dimension (after chromatographic separation) of sample characterization by using a quadrupole GC-MS with negative ion chemical ionization (NICI) and selected ion monitoring (SIM). NICI was the preferred ionization method because of the highly electrophilic nature of these chlorinated pesticides. Mostly [M – Cl]⁻ ions were obtained and sensitivities were ~100-fold higher than those observed with either EI or CI. Also, NICI is transparent to compounds that are not electrophilic, and therefore not ionized, e.g., polynuclear aromatic hydrocarbons. SIM isolates co-eluting analytes with different m/z values, while GC retention times can be used to distinguish isobaric ions, e.g., heptachlor, octachlorostyrene, and dieldrin. However, major components of chlordane still did produce ions that interfered with the monitoring of the hexachloro components of toxaphene. Other interferences arose from fragments of other contaminants and from toxaphene itself. In the latter case, an example is the ^{13}C isotope ion of an [M – HCl]⁻ fragment ion that interferes with the ^{12}C of the [M – Cl]⁻ ion, as they both have the same mass. Choosing specific sets of SIM ions, including monitoring of selected masses within the chlorine isotope patterns, enabled the identification and quantification of the toxaphene while permitting subtraction of the concentrations of interfering compounds.

Sample preparation relied on the specificity provided by NICI and SIM to simplify sample extracts. The fish tissue was spiked with an internal standard, dried with sodium sulfate, Soxhlet extracted with hexane/acetone, and had the lipids removed by deactivated silica gel chromatography. Quantification was based on components with 6–10 chlorine atoms, as these represented the majority of the compounds of toxaphene. The concentrations of the bornane and bornene components were combined because of overlaps of both GC retention times and isotope patterns.

Relative response factors derived from standards were factored into the calculation of final concentrations because NICI is compound specific in terms of its sensitivity.

Results and discussion: The detection limit for total toxaphene was 75 pg (on column). Toxaphene concentrations were compared in composite fish samples from two locations in the Great Lakes: Siskiwit Lake, which is in a remote wilderness on Isle Royale in Lake Superior, and Saginaw Bay in Lake Huron, an area contaminated with organochlorines that is heavily industrialized and has a high population density. The concentrations of toxaphene in fish were 220–290 ng/g at Siskiwit Lake and 510 ng/g at Saginaw Bay. These results were significantly different from those observed for other chlorinated pesticides that occurred at 100-fold higher concentrations in fish from Saginaw Bay when compared to those from Siskiwit Lake. The fact that toxaphene concentrations in the two locations differed by only a factor of two indicates that there was no obvious point source contamination in the generally much more polluted Saginaw Bay region. Instead, a more indiscriminate distribution of toxaphene must have occurred, suggesting that aeolian transport was the likely source of the contaminant in the Great Lakes. Some changes in the relative levels of chlorination, higher levels of 6 Cl species, and reduced amounts of the 8–10 Cl components, were observed for predatory species of fish, implying that there may be metabolism of the toxaphene as it moves through the food chain.

4.3 ENVIRONMENTAL: PHARMACEUTICALS IN SURFACE AND WASTEWATERS

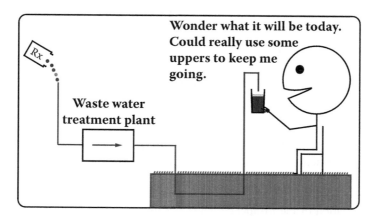

Field: Pharmaceuticals, environmental chemistry.
Techniques: Solid phase extraction, LC-ESI-MS/MS (QqQ), selected reaction monitoring.
Reference: M. Gros, M. Petrovic, and D. Barcelo, Development of a multi-residue analytical methodology based on liquid chromatography-tandem mass spectrometry (LC-MS/MS) for screening and trace level determination of pharmaceuticals in surface and wastewaters, *Talanta* 70:678–690 (2006).

Background: Pharmaceuticals and personal care products are being used in increasing quantities and are now considered "emerging pollutants of concern," partly because of the inefficient removal of small molecules by wastewater treatment plants. While the concentrations of these pollutants in the environment are low, there is a lack of knowledge of the long-term health risks for humans and nontarget species.

Methods: Solid phase extraction, ESI-LC-MS/MS (QqQ), and selected reaction monitoring (SRM) were used to analyze surface waters as well as influent and effluent flows at a water treatment plant. Twenty-nine compounds were analyzed from the following categories: eight analgesics and anti-inflammatories, five lipid regulators and cholesterol-lowering drugs, three psychiatric drugs, one antiulcer compound, three histamine receptor antagonists, five antibiotics, and four β-blockers.

Results and discussion: Representative examples of extraction efficiencies obtained using four different solid phase extraction cartridges (polymeric, C18, and two ion exchange) as well as two pH values for the most effective polymeric cartridge (Oasis HLB) are shown in Figure 4.3. Contaminants were recovered effectively from river water as well as from the influent and effluent flows of a water treatment plant. Recoveries were in the 60–102% (surface water) and 50–116% (wastewater) range, with a few compounds in the 35–50% range. Calibration curves for all compounds had r^2 values of >0.99.

Sensitivity was improved by switching through selected sets of SRM ions at specific times during the LC gradient, rather than monitoring continuously for all compounds. Monitoring a second SRM transition for each analyte increased specificity by ensuring that the ratios of the intensities of the two ions were consistent with those

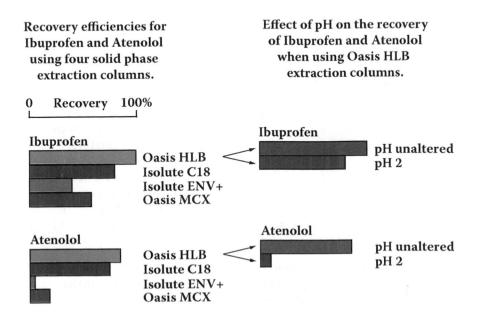

FIGURE 4.3   Examples of the extraction efficiencies for contaminants in wastewater.

observed for the standards. Detection limits were improved when different mobile phase compositions were used for positive or negative ion detection, so the target compounds were divided into two classes instead of attempting to analyze all in one run. Deuterated internal standards were used to improve the accuracy of quantifications. The internal standards were usually effective in compensating for any ion suppression (in the ESI process) attributable to the large amounts of co-extracted material in the samples.

Various members of each of the seven categories of drugs were detected: analgesics and anti-inflammatories (6 of 8), lipid regulators and cholesterol-lowering drugs (4 of 5), psychiatric drugs (1 of 3), antiulcer (0 of 1), histamine receptor antagonists (1 of 3), antibiotics (4 of 5), and β-blockers (3 of 4), giving a total of 19 of the 29 drugs investigated.

Individual contaminant concentrations in river water from Spain ranged from below detection limits (typically <10 ng/L) to <100 ng/L, with occasional spikes for ibuprofen, acetaminophen, and atenolol. Wastewater samples (from a water treatment plant in Croatia) showed the same compounds, but at concentrations up to 100-fold (or higher) more than those observed in river water. The distribution of the average effluent concentrations for each drug category is shown in Figure 4.4. Analgesics and anti-inflammatory drugs were the major group, with acetaminophen having the highest average concentration, constituting 70% of the analgesic and anti-inflammatory category and 36% of all the drugs monitored. Average concentrations in the influent and effluent samples were similar, demonstrating the inefficiency of the water treatment plant for removing small molecules. The concentrations of contaminants were similar to those reported for treatment plants in Spain, Germany, Italy, and the United States.

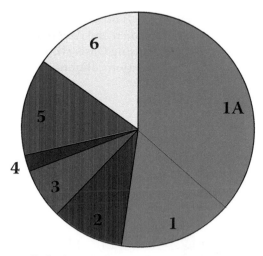

Drug categories

1. **Analgesics, anti-inflammatories**
   A = acetominophen
2. **Lipid regulators, statins**
3. **Psychiactric drugs**
4. **Histamine antagonists**
5. **Antibiotics**
6. **β-blockers**

Relative concentrations of drugs in each category in wastewater treatment plant effluent. Acetominophen was the single largest contributor comprising 70% of the analgesics.

FIGURE 4.4   Drug distribution in wastewater.

4.4 PHARMACOLOGY: LIPITOR METABOLISM

Field: Analytical pharmacology.

Technique: LC-MS/MS with selected reaction monitoring (SRM)

Reference: J. S. Macwan, I. A. Ionita, M. Dostalek, and F. Akhlaghi, Development and validation of a sensitive, simple, and rapid method for simultaneous quantitation of atorvastatin and its acid and lactone metabolites by liquid chromatography-tandem mass spectrometry (LC-MS/MS), *Anal. Bioanal. Chem.* 400:423–433 (2011).

Background: It is well known that hypercholesterolemia is a major risk factor in the progression of atherosclerosis, the major cause of cardiovascular diseases. Statins are widely used to treat hypercholesterolemia. The mechanism of action of these drugs is to reduce the endogenous production of cholesterol by inhibiting 3-hydroxy-3-methylglutaryl coenzyme (HMG-CoA) reductase. Atorvastatin (ATV, Lipitor) is one of the top-selling prescribed oral medications. The only known adverse effect is skeletal muscle toxicity (myopathy) that may be related to the formation of the lactone of the acidic side chain on the molecule.

Methods: The six analytes were the parent drug (ATV), the active *ortho-* and *para-*hydroxy-ATV, and the inactive lactones of each compound. Deuterated (D_5) derivatives of each served as internal standards. The analytes, recovered after protein precipitation of 50 µl plasma, were separated with LC using a solvent gradient, ionized by positive ESI, and detected using selected reaction monitoring of diagnostic (Q1 to Q3) ions on a triple quadrupole mass spectrometer. Method validation with respect to selectivity, sensitivity, accuracy, precision, recovery, and stability was made using guidelines of the U.S. Food and Drug Administration.

Results and discussion: Minimizing and acid:lactone conversion during sample preparation is important if the lactone is, as suggested, a marker for myopathy. For this reason, sodium fluoride was preferred (several anticoagulants were assessed), as it is an esterase inhibitor, and proteins were precipitated using acetonitrile with 0.1% glacial acetic acid. Conditions for LC were developed so that all analytes (i.e., the parent drug and all metabolites) were baseline separated. This was essential because the precursor-to-product ion transitions of the *ortho-* and *para-* analytes are the same, and also because any in-source fragmentations might produce interfering ions. The major product ions from the ATV precursor ion, $[M+H]^+$, were from the neutral loss of the phenylaminocarbonyl moiety. There were phenylamino group losses from the acid and lactone metabolites, respectively (Table 4.1). The calibration curves for each analyte were linear over a concentration range of 0.05–100 ng/ml ($r^2 > 0.997$). The lower limit of quantification (LLOQ) for all analytes was 0.05 ng/ml (S/N = 5:1). This was important because the *para-* compounds occur at ~0.1 of the concentration of the *ortho-* compounds. Compound recoveries were in the 89–111% range. Data were obtained on the stability of ATV and metabolites in blood plasma under various experimental conditions, including keeping extracts at 4 °C for 24 h, repeated freeze-thawing, and a 3-month storage period at –80 °C. The samples were stable in all cases.

The method was applied in a clinical study to quantify ATV and its metabolites over a 12 h post-dosing period in blood samples from kidney transplant patients receiving ATV and immunosuppressive drugs (Table 4.1). The main use of the technique is the (previously unattainable) simultaneous determination of the concentrations of the parent drug and all its primary metabolites for pharmacokinetic studies. Although the mechanism of ATV-related myopathy is not known, the lactone-to-parent concentration ratio may have utility as a diagnostic marker.

TABLE 4.1

Data on the Determination of Atorvastatin and Its Metabolites in Plasma from Kidney Transplant Patients

Structures of ATV and its metabolites	Transition monitored in MRM	plasma ATV conc 4 h after 10 mg dose ng/mL	mean plasma metab: ATV ratios n = 9
ATV	559 > 440	2.5	
ATV-lac	541 > 448	3.6	0.8–2
o-OH-ATV	575 > 440	1.5	0.4–0.6
o-OH-ATV-lac	557 > 448	3.6	2–4.2
p-OH-ATV	575 > 440	0.1	0–0.2
p-OH-ATV-lac	575 > 448	0.1	1–1.6

4.5 NEW TECHNIQUES: PAPER SPRAY OF PHARMACEUTICALS

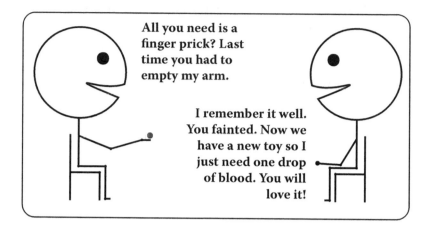

Field: Method development.

Techniques: Paper spray, MS/MS.

References: H. Wang, J. Liu, R. G. Cooks, and Z. Ouyang, Paper spray for direct analysis of complex mixtures using mass spectrometry, *Angew. Chem. (Int. Ed. Engl.)* 49:877–880, (2010). N. E. Manicke, P. Abu-Rabie, N. Spooner, Z. Ouyang, and R. G. Cooks, Quantitative analysis of therapeutic drugs in dried blood spot samples by paper spray mass spectrometry: An avenue to therapeutic drug monitoring, *J. Am. Soc. Mass Spectrom.* 22:1501–1507, (2011).

Background: Mass spectrometry has the potential to become an important tool in therapeutic drug monitoring, especially for drugs with narrow effective concentration ranges. Current methodologies are limited by the need for relatively large (100 μl or more) amounts of blood, complex sample workups, and analysis protocols, e.g., drug extraction and UPLC-ESI-MS/MS.

Methods: Ambient ionization methods, of which there are now over 20, e.g., desorption electrospray ionization (DESI), desorption atmospheric pressure chemical ionization (DAPC), desorption atmospheric pressure photo-ionization (DAPPI), and direct analysis in real time (DART), are now joined by "paper spray," a method where ESI is initiated at the pointed tip of a piece of filter paper. A drop of blood (~15 μl) is dried on the paper, and then the paper is moistened with ~25 μl of a solvent suited to both the extraction of the analytes from the blood and the ESI process (e.g., 90% methanol:10% water with either 100 ppm acetic acid or 200 ppm sodium acetate). When the paper is exposed to high voltage (3–5 kV) while held close (~5 mm) to the entrance of the mass analyzer, a spray (similar to electrospray) is induced at the tip of the paper as capillary action carries extracted compounds through the paper (Figure 4.5). The spray is maintained for 30–90 s at a flow rate comparable to that used in nano-electrospray.

Results and discussion: The technique was evaluated for a number of drugs, and experiments were conducted to determine the limits of detection, reproducibility,

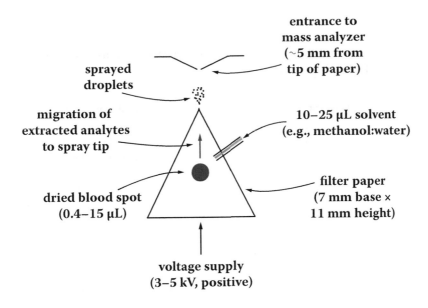

FIGURE 4.5 Paper spray source.

and whether there were matrix effects. Highly reproducible data were generated for many drugs, although there were major differences in sensitivity (Table 4.2). Deuterated internal standards were dried onto the paper first, directly below where the blood spot would be placed. There were significant differences in detection limits, from <0.1 to >100 ng/ml. The lower limits of detection for mildly basic and hydrophobic drugs were <1 ng/ml.

In another application, paper used to wipe a surface contaminated with heroin and cocaine was cut to provide the necessary emission point; both drugs were detected. Paper spray was also shown to ionize various other compounds effectively, including methyl violet (a dye), phosphatidylcholine (a lipid), angiotensin (a peptide), and the protein cytochrome C.

Paper spray methodology requires no modification of the mass spectrometer. The technique is carried out most effectively on MS/MS instruments (a triple quadrupole was used in these papers) as MS spectra tend to include many miscellaneous co-extracted compounds. CID with detection of the product ion spectra or operation in SRM mode enables both identification and quantification of the analytes. Paper spray is most effective in positive ion mode, although it is possible to use negative mode if the ESI voltage and solvent composition are altered to prevent a discharge.

Paper spray is a novel method for assessing therapeutic concentrations for many drugs and could be applied in clinical settings. With appropriate hardware and software, the technique could be automated, providing rapid, inexpensive drug screening/assays that could be carried out by personnel without training in mass spectrometry.

TABLE 4.2

Detection Limits and Reproducibility Data for Drugs Analyzed in Dried Blood Spots Using Paper Spray

Drug	Precursor Ion	Product Ion	LLOD ng/ml[a]	Experiments and Results[b]
Acetaminophen	152, [M + H]⁺	110	250	
Amitriptyline	278, [M + H]⁺	233		For 5 different blood samples signals from a 1 ng/ml sample were always >5× the blank[c] For 6 concentrations between 0.89 and 443 ng/ml imprecision was 8–22% (RSD) ($^2H_6$ IS[d])
Benzethonium[e]	412, [M]⁺	320	0.02	
Citalopram	325, [M + H]⁺	109		Calibration curves for 5 different human blood samples had slopes that differed by only 1.3%, showing no relative matrix effect ($^2H_4$ IS) Precision varied from 19% (RSD) at 1 ng/ml to 2% at 500 ng/ml
Dextrorphan	258, [M + H]⁺	157	0.6	
Ethoxycoumarin[f]	191, [M + H]⁺	163	1	
Ibuprofen	205, [M − H]⁻	161	500	
Imatinib	494, [M + H]⁺	394		Calibration curve between 60 and 4000 ng/ml gave r² of >0.999 using a $^2H_8$ IS
Paclitaxel	876, [M + Na]⁺	308	15	
Proguanil	254, [M + H]⁺	170	0.08	
Simvastatin	441, [M + Na]⁺	325	50	
Sitamaquine	344, [M + H]⁺	271		For concentration from 5–1,000 ng/ml imprecision was always <4% (RSD)
Sunitinib	399, [M + H]⁺	283	0.25	Calibration curve was linear ($^2H_{10}$ IS) For 8 (×2 sets) comparison samples: for concentrations of 2.9 and 250 ng/ml, the observed results = 2.6 and 263 ng/ml, respectively For 5 different blood samples the signal from a 1 ng/ml sample was always >5× the blank
Telmisartan	515, [M + H]⁺	276	0.3	For 5 different blood samples the signal from a 1 ng/ml sample was always >5× the blank
Verapamil	455, [M + H]⁺	165	0.75	

[a] LLOD, ng/ml = lower limit of detection.

[b] All variations reported are within guidelines for analytical performance set by the Food and Drug Administration and the Clinical Laboratory Improvements Amendments.

[c] See p. 151.

[d] ^2H_x IS = extent of deuteration of the internal standard (IS) for the drug used, i.e., $^2H_4$ or $^2H_6$.

[e] Quaternary ammonium compound, [M]⁺ observed.

[f] Model compound (not a drug).

4.6   PETROLEOMICS: CRUDE OIL CHARACTERIZATION

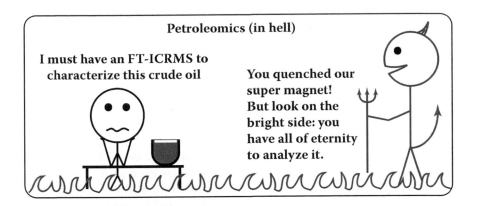

Field: Petroleomics.

Technique: Fourier transform ion cyclotron resonance mass spectrometer (FT-ICRMS).

Reference: A. G. Marshall and R. P. Rogers, Petroleomics: Chemistry of the underworld, *Proc. Natl. Acad. Sci. USA* 105:18090–18095 (2008).

Background: Crude oil is an increasingly valuable commodity. As supplies of low-sulfur light, "sweet" crudes are depleted, characterization of the composition of the heavier, more acidic, higher sulfur oils is becoming critical for developing processing methods, as well as in provenance and point-of-origin determinations.

Methods: Crude oil is a highly complex mixture composed of thousands of components, thus conventional chromatography is of no value. The majority (~90%) of the constituents are nonpolar and thus inaccessible to ESI. Atmospheric pressure photo-ionization (APPI) can be a useful ionization method, because many of the components include aromatic groups. One difficulty with APPI is that both $[M]^{+\bullet}$ and $[M + H]^{+}$ are formed, meaning that the nitrogen rule cannot be applied, and also that the $[M(^{13}C_1)]^{+\bullet}$ and $[M + H]^{+}$ ions from the same analyte must be resolved (mass difference = 4.5 mDa). Furthermore, the sulfur content is an important factor in refining strategies, and high mass resolution is an appropriate method for providing information on the presence and distribution of sulfur. For instance, only 3.4 mDa separate two compounds that have the same nominal mass (isobaric) but empirical formulae that differ by C_3 and SH_4 (36 Da nominal), respectively. The very high resolution (~400,000) and mass accuracy (<0.3 ppm) available with FT-ICRMS enable the separation and mass measurement of the 4.5 and 3.4 mDa differentials and provide an approach to characterize oils and obtain empirical formulae for many components in the mixtures.

Results and discussion: Analysis of crude oil by FT-ICRMS can generate lists that may contain 60,000 items. Such lists are impractical, and the data must be reprocessed to categorize their information content. The first set of criteria are the numbers and types of heteroatoms present, e.g., N_1, O_1, N_2, NO, O_2, NS, and O_3. Within each heteroatom-based group there are various homologous series that can be sorted, first by their double bond equivalents (DBEs) and then by their carbon number. This processing provides a nested set of results, i.e., heteroatom > DBE > carbon number. The homologous sets of compounds can also be categorized according to their substituent groups, and in particular by the numbers of methylenes ($-CH_2$) present using the so-called Kendrick scale, in which CH_2 is defined as 14.0 instead of the conventional 14.0565. Redefining the mass scale in this way means that any given set of homologous compounds will carry a mass defect that is characteristic of that

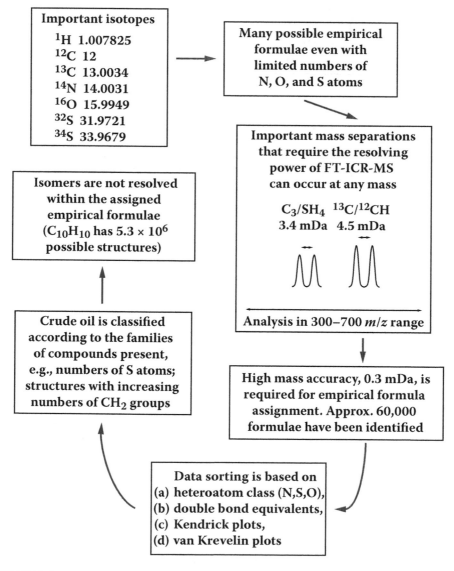

FIGURE 4.6   Processing of FT-ICRMS data in petroleomics.

structure and can be followed across the mass range over which data were collected. Another method to categorize the data is to generate van Krevelen plots which are constructed by taking the hydrogen/carbon (H/C) ratios (y-axis) and then plotting them against an x-axis that consists of heteroatom/carbon ratios, e.g., nitrogen/carbon (N/C), sulfur/carbon (S/C) or oxygen/carbon (O/C).

These methods of data categorization are ways of creating visual illustrations of the classes of compounds that can be used to classify crude oils according to their chemical properties. One aspect that cannot currently be addressed is the matter of isomeric structures, e.g., $C_{14}H_{10}$ has 5.3×10^6 possible structures. The sequence of events in the processing of FT-ICRMS data from a crude oil sample is illustrated in Figure 4.6. Quantification is difficult because ionization efficiencies (instrument response) are a function of compound structure with both APPI and ESI.

To date, FT-ICRMS analysis has yielded more than 60,000 monoisotopic compositions (^{12}C, ^{1}H, ^{14}N, ^{16}O, ^{32}S) using a combination of APPI, ESI, and positive and negative ion detection. These data have led to the construction of a *petroleome database*. Other information about crude oils in this database includes geographic origin, density, total acid number, corrosivity, and distillation profiles (sorted by aromatic class, etc.). All data are searchable and can simplify characterization and provenance determinations. Similar databases can be envisaged for other highly complex mixtures, such as the components of coal and biofuels as well as humic and fulvic acids.

4.7 METABOLOMICS: DISEASE MARKERS FOR A TROPICAL DISEASE

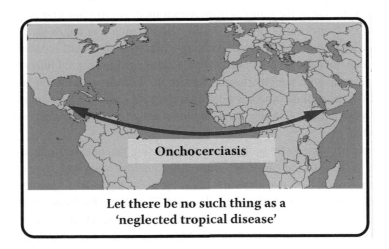

Fields: Metabolomics, disease markers.

Techniques: LC-MS, statistical analysis.

Reference: J. R. Denery, A. A. K. Nunes, M. S. Hixon, T. J. Dickerson, and K. D. Janda, Metabolomics-based discovery of diagnostic biomarkers for onchocerciases, *PLoS Negl. Trop. Dis.*, 4:e834 (2010).

Background: Onchocerciasis (river blindness), caused by the nematode *Onchocerca volvulus*, is classified (by the World Health Organization) as a "neglected tropical disease," even though it seriously affects 37 million sufferers and places 95 million people at risk in countries from Central America across Africa to Yemen. An effective marker in blood would help eliminate current diagnostic problems, which include poor sensitivity, variations in infective species, and immunogenicity, all of which hinder elimination programs.

Methods: Stored samples (both plasma and serum) from prior studies (1986–2008) in Ghana, Cameroon, Guatemala, Liberia, India, and the United States were used. Samples were analyzed, after protein precipitation, by LC-ESI-TOF. Specific analytes underwent additional studies using QTOF for MS/MS and FT-ICRMS for accurate mass measurement. Results were processed using principal component analysis (PCA) and machine learning algorithms.

Results and discussion: Over 2,000 data components were registered initially using the evaluation software. The data set was reduced with various filters and probability stringency actions to a subset of 14 compounds that could be evaluated as potential markers. Partial identification of the 14 compounds (by MS/MS and accurate mass measurement) showed that 10 were lipidic (molecular mass 240–530 Da), and the others were peptides (~1 kDa). Notably, the concentrations of all but one of the candidate biomarkers were depressed (1.5- to 5.6-fold) in the pathological samples when compared to controls. Results were consistent in both plasma and serum.

PCA comparison of the samples from Africa successfully (100%) segregated diseased (n = 55) from control (n = 18) states. The situation was not as clear when all geographic regions were compared. Although there was an overall separation of the disease group (including from samples with a closely related filarial infection), there was some overlap with samples positive for Chagas disease and leishmaniasis. When patients from Guatemala, who were being actively treated and where biopsies (24%) showed no live parasites, were compared with controls, the two groups could not be distinguished.

A quantitative diagnostic statistic was sought using machine learning algorithms, where the value obtained can vary from 0.5 for random chance to 1.0 for a perfect match. Analysis of the mass spectral data by these algorithms generated values between 0.87 and 1 for the African samples and 0.77 and 0.95 for the entire sample set.

The fact that the diseased and control samples from Guatemala could not be differentiated requires further investigation, but suggests that the candidate biomarkers may indicate active infection. That there was some cross-identification with other diseases shows that additional refinement of the biomarker search is required, e.g., collection of negative ion spectra and use of alternative chromatographic techniques. The cross-identification suggests that the metabolism of the host is affected by other diseases; thus, it may be possible to find other biomarkers, including for the differential diagnosis of Chagas disease and leishmania infections.

4.8   METABOLOMICS: CHEMICAL DEFENSE

> Just keep eating this red algae. It's an invasive species and we have to get rid of it.
>
> I'm trying, but if I get anywhere near where you have been eating it tastes like @$*&%.

Field: Metabolomics in invasive species.

Techniques: LC-MS, GC-MS, statistical analysis.

Reference: G. M. Nylund, F. Weinberger, M. Rempt, and G. Pohnert, Metabolic assessment of induced and activated chemical defence in the invasive red alga *Gracilaria vermiculophylla, PLoS One* 6:e29359 (2011).

Background: Much of chemical ecology has been focusing on chemical defenses in terrestrial plant–herbivore interactions, mostly ignoring the marine environment. It has been argued that plants with defense mechanisms succeed because their likely consumers are "generalist" herbivores that graze on a variety of species, and therefore have the ability to switch food sources and avoid species with chemical defenses. Plant defense mechanisms may be constitutive, activated, or induced, each with different timescales. Palatability measurements are used to assess the effectiveness of chemical defenses. Until recently, attempts to identify active chemical deterrents relied on chromatographic separation of individual candidate compounds that were then used in feeding trials. Metabolomics, with statistical analysis, enables comprehensive evaluation of chemical defenses, including the potential for revealing synergistic processes.

The red alga *Gracilaria vermiculophylla* is native to the Northwest Pacific but has become an invasive species in the Eastern Atlantic and Baltic Sea. Observations of this alga suggest that it has a chemical defense mechanism, a condition that is likely to enhance its ability to invade new habitats, because it is a low-preference food source for generalist herbivores.

Methods: Experiments were undertaken to determine whether the metabolic profiles of the alga changed when its tissues were mechanically damaged by grinding or mutilation with forceps. Other tests investigated whether intact tissues were affected when neighboring tissues were injured, and whether exposure to extracts of damaged tissue could alter the metabolic profile of the normal tissue. The isopod *Idotea baltica* was used in two-way feeding assays where this herbivore was able to feed on media to which extracts of intact or ground-up *G. vermiculophylla* had been added. Once prospective repellent compounds, including prostaglandins and hydroxyeicosatetraenoic (HETE) acids, were identified, feeding assays were carried out using media supplemented with the compounds of interest.

Methanol extracts of selected tissues were analyzed by UPLC-ESI-QTOFMS (negative ion mode). Derivatized (methoxyamination and trimethylsilylation) samples were also analyzed by GC-TOF-MS. Comparison of unknown spectra with co-injection standards and NMR-separated fractions were used for analyte identification. Statistical analysis included canonical analysis of principal coordinates (an extension of principal component analysis) and permutational multivariate analysis of variance.

Results and discussion: Results of the LC-MS results revealed that 11 metabolites were upregulated and 3 were downregulated in the damaged algal tissue. The GC-MS data showed eight upregulated compounds. The predominant species, 8-HETE and 7,8-di-HETE, prostaglandins PGE_2, 15-keto-PGE_2 and PGA_2, and an unknown compound, were 70–400 times upregulated in the damaged tissue when compared with normal tissue. In addition, there were significant increases in PGE_2, 15-keto-PGE_2, and 7,8-di-HETE in response to herbivore grazing on intact tissue. Metabolic changes were also observed in intact tissue that was exposed to damaged tissue extract. The herbivores in the feeding assays showed a preference for the control samples, i.e., those exposed to the extracts of the intact alga. Other trials, where individual identified compounds were added to the food, suggested that PGA_2 was the most important defense agent.

The defense mechanisms of *G. vermiculophylla* seem to involve the arachidonic acid pathway that can lead to prostoglandins and hydroxylated unsaturated fatty acid, such as members of the HETE family. Addition of ^{2}H-labeled arachidonic acid to wounded alga showed that labeled fatty acids were generated rapidly. Interestingly, these results suggested that the enzymatic processes necessary for the activation of the defense mechanism remained viable, despite the damage to the tissues. The metabolomic assay used validated the existence of the chemical defenses and characterized the mechanisms and biochemical pathways involved.

4.9   LIPIDOMICS: CORONARY ARTERY DISEASE

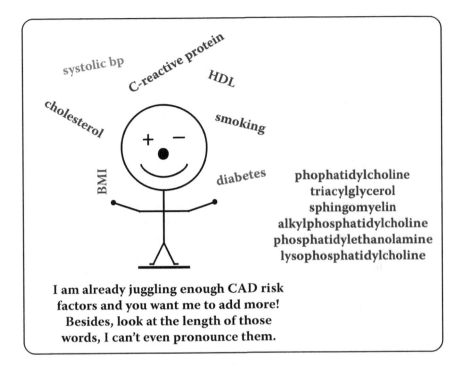

I am already juggling enough CAD risk
factors and you want me to add more!
Besides, look at the length of those
words, I can't even pronounce them.

Field: Differential diagnosis of coronary artery disease.

Techniques: LC-ESI-MS/MS, selected reaction monitoring (SRM), advanced multivariate classification techniques.

Reference: P. J. Meikle, G. Wong, D. Tsorotes, C. K. Barlow, J. M. Weit, M. J. Christopher, G. L. MacIntosh, B. Goudey, L. Stern, A. Kowalcyk, I. Haviv, A. J. White, A. M. Dart, S. J. Duffy, G. L. Jennings, and B. A. Kingwell, Plasma lipidomics analysis of stable and unstable coronary artery disease, *Arterioscler. Thromb. Vasc. Biol.* 31:2723–2732 (2011).

Background: It is well known that acute coronary syndromes, e.g., unstable angina and myocardial infarction, result from the disruption of atherosclerotic plaques followed by thrombosis. The roles of plasma cholesterol levels and the ratio of low- to high-density lipoproteins have long been recognized in coronary artery disease (CAD), as have other physiological measurements, including blood pressure and calcification of coronary arteries. Risk assessment also includes lifestyle factors, such as smoking status and body mass index. The objective of this paper was to investigate the potential use of plasma lipid profiles in determining the risk of formation of unstable plaques and poor health outcomes.

Methods: Plasma samples were obtained from three matched populations, one with stable CAD (n = 60), another with unstable CAD (n = 80), and a control group (n = 80). Ten µl aliquots of randomized plasma samples (treated with butylhydroxytoluene, an antioxidant) were extracted with chloroform:methanol (2:1 v/v) to obtain a lipid-rich fraction. Lipids were analyzed by LC-MS/MS on a triple quadrupole mass spectrometer. Precursor ion and neutral loss scans were carried out for some 305 lipid species from >20 known lipid groups, including ceramides, G_{M3} gangliosides, sphingomyelins, phospholipids, and cholesterol esters (CEs). SRM analyses were made on samples from the 220 individuals. Data analysis was based on both individual lipids and total quantities for each lipid class.

Multiple statistical analyses included linear regressions, χ^2 t, and Mann-Whitney U tests. Multivariate classification models were created for the statistical evaluation of risk using a feature selection program (ReliefF) coupled with an L2-regularized logistic regression-based classifier, followed by using a three-fold cross-validation program to assess the performance of the models developed.

Results and discussion: Specific classes of lipids were associated with stable (13) and unstable (5) CAD. Even when an entire class was not obviously related to the disease, individual lipids within that class could be identified as discriminatory for unstable CAD, suggesting their involvement in the development of the syndrome. For example, lipid species with shorter chain lengths that were saturated or had a limited number of double bonds (particularly CEs 14:0, 16:1, and 16:2) showed a negative correlation with unstable CAD (vs. stable CAD), whereas CEs and diacylglycerol species with longer chain lengths (e.g., CEs 20:3 and 20:4) were positively related to stable CAD (vs. control group). Unsaturation was linked to stable CAD, notably phoshatidylinositols containing a 20:4 chain.

Statistical evaluation enabled reduction of the number of lipids used in the risk assessment from 305 to 105. Risks models used were lipids alone, traditional risk factors, and the combination of lipids and risk factors. The results were expressed using the C-statistic (which varies from 0.5 for random association to 1.0 for a perfect

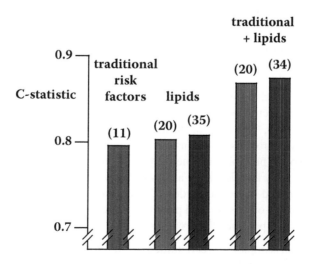

FIGURE 4.7   Risk models for coronary artery disease.

outcome) as a measure of risk, so that the effectiveness of the models could be compared. When the categories were considered independently, C values for the discrimination between stable and unstable CAD were 0.796 for traditional risk factors (11) and 0.802 for lipids (20). When the two categories were used together the C-statistic increased to 0.869 (20 features) (Figure 4.7). It was noticeable that the C-statistic appeared to reach a plateau, at which point there were limited gains from increasing the number of risk factors, e.g., from 20–35 for lipids and 20–34 for traditional risks + lipids (Figure 4.7).

The top six traditional risks were C-reactive protein, systolic blood pressure, smoking, cholesterol, CAD history, and body mass index. The top six lipid species involved were phosphatidylcholine (PC) 34:5, PC 18:1/18:3, triglyceride 14:0/16:0/18:2, PC 28:0, sphingomyelin 18:2, and the odd carbon PC 33:3. When the two sets of risk factors were combined, the top six were PC 34:5, PC 18:1/18:3, C-reactive protein, triglyceride 14:0/16:0/18:2, PC 28:0, and systolic blood pressure. The results argue that models that include both lipids and traditional risk factors provide statistically more significant classification of unstable vs. stable CAD compared with models based on only traditional factors.

4.10 PROTEOMICS: PROTEIN IDENTIFICATION IN A PAINTING

I hate to tell you – but all that has been keeping us stuck in this fresco for 100 years is a bit of egg glue.

Field: Proteomics, protein i.d.
Technique: Electrospray QTOF with DDA for peptide sequencing.
Reference: A. Chambery, A. Di Maro, C. Sanges, V. Severino, M. Tarantino, A. Lamberti, A. Parente, and P. Arcari, Improved procedure for protein binder analysis in mural painting by LC-ESI/Q-q-TOF mass spectrometry: Detection

of different milk species by casein proteotypic peptides, *Anal. Bioanal. Chem.* 395:2281–2291 (2009).

Background: Understanding the biochemistry of objects of cultural heritage contributes to our historical knowledge and provides information for the conservation of art. This paper reports on the identification of proteinaceous binders in a mural in the St. Dimitar Cathedral in Vidin, Bulgaria, that was probably painted by Kiril Jeliaskov in the early twentieth century.

Methods: Four scrapings were obtained during a restoration effort. Approximately 50 mg of the painting surface, corresponding to 0.1 cm², was excised and resuspended (with sonication) in ammonium bicarbonate buffer. The samples were reduced with dithiothreitol, alkylated with iodoacetamide, and immediately treated with trypsin (overnight, at 37°C). The digests were dried, resuspended in 0.1% formic acid, and centrifuged to remove particulate matter. LC-MS was carried out with a symmetry C_{18} column (10 cm × 300 μm i.d.) using a solvent gradient of 5 to 55% acetonitrile with 0.1% formic acid (flow rate: 5 μl/min). MS/MS spectra (*m/z* 50–1,600 Da) were collected on doubly and triply charged ions in data-dependent analysis (DDA) mode. Mass fingerprint data on the molecular masses of the peptides were also collected. Spectra were centroided, deisotoped, and charge state reduced to provide accurate mass values for the monoisotopic masses of the peptides and their fragments. The data were searched against the Swiss-Prot database for protein identification (Figure 4.8).

Homemade tempera paints were prepared according to old recipes dating from 1859 (egg based) and 1894 (milk based); in the latter, sheep, goat, and bovine milk were used. Mixtures of milk and egg tempera were placed on microscope slides and dried at room temperature. It was shown that ~1 μg of egg protein was required to give good quality MS/MS spectra, appropriate for protein identification.

Results and discussion: The MS/MS spectra from the samples of the mural indicated the presence of ovotransferin, ovalbumin, and lysozyme from egg white, and the egg yolk protein vitellogenin. The average number of peptides identified from these four proteins was 8, 6, 2, and 26, respectively. The results suggested that the simple sample workup effectively preserved egg-based proteins, including vitellogenin, from the scraping.

The mural samples also contained caseins, alpha S1 (six peptides) and beta (five peptides), indicating that milk proteins had been used at some point, either in the original painting or in an undocumented restoration effort. The database searches identified the alpha-casein as originating from sheep despite the 97% similarity in the sequence of the protein in sheep and goat.

The homemade temperas (sheep, goat, cow) were also investigated to establish if the species from which the milk was obtained could be determined. Differences between the alpha S1 casein from the three species are shown in Table 4.3. There are multiple (23) amino acid substitutions that can be used to distinguish bovine milk from sheep and goat milk. The latter two species are harder to separate, as there are only five amino acid substitutions. The elimination of K at position 22 in goat milk (K to N) removes a trypsin cleavage site; thus, the 23–37 peptide chain

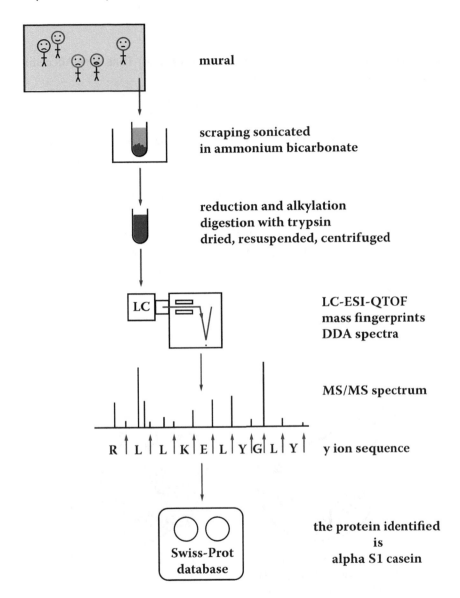

mural

scraping sonicated
in ammonium bicarbonate

reduction and alkylation
digestion with trypsin
dried, resuspended, centrifuged

LC-ESI-QTOF
mass fingerprints
DDA spectra

MS/MS spectrum

R L L K E L Y G L Y   y ion sequence

Swiss-Prot
database

the protein identified
is
alpha S1 casein

FIGURE 4.8   Protein identification in a mural paint sample.

is not observed for goat milk. In addition, the 209–214 chain has different masses
in sheep (760 Da) and goat (748 Da), resulting from an I-to-T transition. Both differences were observed using peptide fingerprint analysis, confirming the ability to
distinguish sheep and goat milk.

The results confirm that species-specific determination of proteins in archival samples is possible. Egg proteins, including yolk protein, were identified, as were proteins
from sheep milk; the latter were unexpected. It is suggested that it is feasible to build a
database of proteotypic peptides to more readily identify binders used in paints.

TABLE 4.3

Positions in Alpha S1 Casein (214 Amino Acid Protein) That Vary According to Species (Sheep, Goat, and Cow)

Species	Amino acid position													
	22	24	27	28	32	39	48	51	52	64	72	76	78	80
Sheep	K	Q	S	P	L	V	R	N	I	I	A	K	G	S
Goat	N	R	S	P	P	V	R	N	I	T	A	K	G	S
Bovine	K	Q	P	Q	L	F	G	K	V	T	I	E	E	I
	81	85	120	129	134	142	143	145	148	152	163	182	207	209
Sheep	A	Y	N	K	Q	N	P	H	Q	A	Q	L	G	I
Goat	A	Y	N	K	Q	N	P	H	Q	A	Q	L	G	T
Bovine	V	H	K	N	R	I	H	Q	E	G	E	V	E	T

Note: Locations where sheep and goat differ are in red. Single letter amino acid designations are used.

4.11   PROTEOMICS: PROTEIN IDENTIFICATION AND METASTASIS

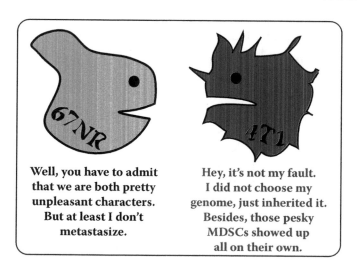

Field: Characterization of specific cell type proteomes.

Methods: Label-free mass spectrometry, shotgun proteomics.

Reference: A. M. Boutte, W. H. McDonald, Y. Shyr, and P. C. Lin, Characterization of the MDSC proteome associated with metastatic murine mammary tumor using label-free mass spectrometry and shotgun proteomics, *PLoS One*, 6:e22446 (2011).

Background: Untreatable metastasis, rather than the primary tumor, is the cause of mortality in breast cancer. Myeloid-derived suppressor cells (MDSCs) are hematopoetic cells that home specifically to the tumors and have a major role in tumor invasion and metastasis and the development of resistance to chemotherapy. MDSCs proliferate in response to tumors and accumulate in the spleen, from which they can be isolated using their Gr1 and CD11b surface markers. The objective was to use label-free mass spectrometry and shotgun proteomics to characterize MDSCs that associate with two mouse cell lines derived from the same tumor, one from the primary tumor (67NR) and the other from cells that have already metastasized to various organs (4T1). Spectral counting, for quantification, and protein network analysis were used to search for MDSC biomarkers characteristic to metastasis.

Methods: Mice were injected with either 67NR or 4T1 cells and permitted to grow for 25 days, after which spleens were harvested and Gr1+/CD11+ MDSCs collected on magnetic antibody beads. Purified cells were counted and lysed in urea buffer, proteins were precipitated and digested (trypsin). Components of the resulting peptide mixtures were isolated on a precolumn packed with C_{18}-type reversed-phase resin followed by a strong cation exchange resin. Pulses of peptides were released (using ammonium acetate) onto a C_{18} column (followed by gradient elution) that was connected to the nanospray ESI source of an LTQ-type ion trap operated in tandem mode in a data-dependent manner (using dynamic exclusion). Mass spectra were analyzed using a mouse database, and proteins were identified with a minimum of two distinct peptides. After statistical analysis, gene ontology, and pathway analysis, proteins (excluding those from immunoglobulin and hemoglobin) were separated into four groups: unique to 67NR, decreased in 4T1, increased in 4T1, and unique to 4T1.

Results and discussion: Of 2,814 identified MDSC proteins 43 were exclusive to 67NR, 153 were exclusive to 4T1, and the rest (2,618) were shared. The shared (2,618) proteins were quantified by spectral counting and searched for up- and downregulation. Using one standard deviation as the threshold showed that 364 proteins were increased and 367 decreased in the 4T1 MDSCs, compared to the 67NR MDSCs. The proteins were primarily derived from the cytosol (34%) and nucleus (26%). Other substantial sources of the proteins were membranes, mitochondria, endoplasmic reticulum, and Golgi apparatus.

The metabolic pathways to which the proteins belong were different, depending on whether the MDSCs came from the 67NR- or 4T1-inoculated mice. The 67NR-derived proteins were involved in basic cellular functions, such as mitosis, nuclear division, and cell cycle regulation, and, notably, downregulation of cell proliferation. Other overrepresented 67NR-MDSC proteins were associated with leukotriene and alkene metabolism. Upregulated proteins in 4T1-MDSC were associated with a wide variety of processes, including energy generation (respiration and the citric acid cycle) and amino acid metabolism. Glutathione and γ-glutamyl transferases were also observed; these enzymes are involved in the control of cell differentiation, and therefore potentially in metastasis. Other upregulated enzymes appear to be involved with chemotherapeutic resistance. There were also proteins associated with processes that include vesicular transport, extracellular-receptor interaction, angiogenesis, and metastatic specific proteins such as colony-stimulating factor receptor 1, pro-platelet basic protein, and arginase 1.

It was concluded that MSDCs are correlated with metastatic potential. The results suggest that the identification of proteins expressed by MDSCs that are unique to metastatic conditions may lead to biomarkers and open possibilities for therapeutic interventions.

4.12   PROTEOMICS: NONCOVALENT INTERACTIONS

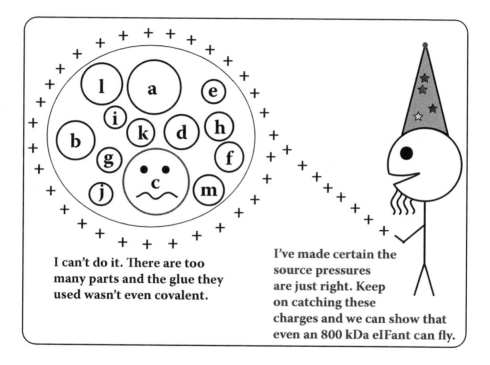

I can't do it. There are too many parts and the glue they used wasn't even covalent.

I've made certain the source pressures are just right. Keep on catching these charges and we can show that even an 800 kDa eIFant can fly.

Field: Proteomics, noncovalent interaction, structural determination, PTM.

Technique: Electrospray for sequencing, post-translational modification location, and noncovalent interactions.

Reference: E. Damoc, C. S. Fraser, M. Zhou, H. Videler, G. L. Mayeur, J. W. B. Hershey, J. A. Doudna, C. V. Robinson, and J. A. Leary, Structural characterization of the human eukaryotic initiation factor complex by mass spectrometry, *Mol. Cell. Proteomics* 6:1135–1146 (2007).

Background: The assembly of the ribosomal complex depends on the presence of proteins called eukaryotic initiation factors (eIFs). In mammals the largest of these is eIF3, an ~800 kDa noncovalent complex comprising of 13 stoichiometric nonidentical subunits ranging in mass from 25 to 167 kDa. This paper describes the isolation and characterization of human eIF3 obtained from HeLa cells (an "immortal" cell line derived from cervical cancer cells from Henrietta Lacks in 1951).

Methods: Initial experiments were to determine which subunits of the complex could be phosphorylated through *in vivo* incubation. The cell lysate fraction containing eIF3 was recovered using differential centrifugation and size exclusion chromatography. The resulting fraction was subjected to reduction, alkylation, and trypsin digestion with recovery of the phosphopeptides by gallium (III) or titanium dioxide enrichment. Nano-LC-MS^2 (or MS^3) was carried out on a linear ion trap FT-ICRMS system for accurate mass determination of the protonated molecules, and either the linear trap or the FT-ICRMS for the detection of the CID fragments. MS^3 was used to trace the presence of phosphopeptides. An ion of interest was selected from the MS 1 spectrum and MS 2 used to search for the CID-induced neutral loss of phosphoric acid (*m/z* 98, 49, or 32.7 for the 1$^+$, 2$^+$, and 3$^+$ states, respectively). Subsequent sequencing of the dephosphorylated species was undertaken using MS^3 after isolation of the dephosphorylated ion and a second CID experiment. Loss of phosphoric acid from a serine resulted in the formation of dehydrolanine, the presence and location of which was determined during sequencing. This process is shown on the left-hand side of Figure 4.9.

Results and discussion: All 13 subunits of eIF3 were identified with an average 79% coverage. Phosphorylation was observed at 29 sites, primarily on subunits a, b, c, and j. The PCI domain, commonly found in protein complexes, was shown to be a constituent of subunits a, c, e, k, l, and m. The locations of the phosphorylations suggest that they may play both internal structural and regulatory roles. The former occurs, for example, when a portion of a protein (or a protein complex) is stabilized by the ionic bonds formed by the alignment of a sequence of negatively charged phosphates with a complementary set of positively charged lysines/arginines. An example of the regulatory function of a phosphate is when serine-46 on eIF3f is phosphorylated during apoptosis.

The molecular mass of the intact eIF3 complex (QTOF instrument) was 804 kDa (right-hand side of Figure 4.9), which was higher than the 795.1 kDa that is the sum of the calculated masses for all the components (including PTM) of the complex. The difference (~1%) may be accounted for by the presence of incompletely removed buffers or solvents and/or by instrumental inaccuracies at these high masses. Furthermore, the variation (~9 kDa) cannot be accounted for by the presence of a second copy of any subunit, the smallest of which has a mass of 25 kDa. Collisional activation of the intact complex promotes the loss of specific subunits, in particular h, l, k, and m, indicating that these units may be located on the periphery of the structure (Figure 4.9, right). In addition, the majority of the phosphorylation sites are on other subunits (a, b, c, and j), suggesting that these subunits form the core of the complex. Ionic interactions of the acidic phosphates with basic groups may be involved in the stabilization of the inner section of eIF3.

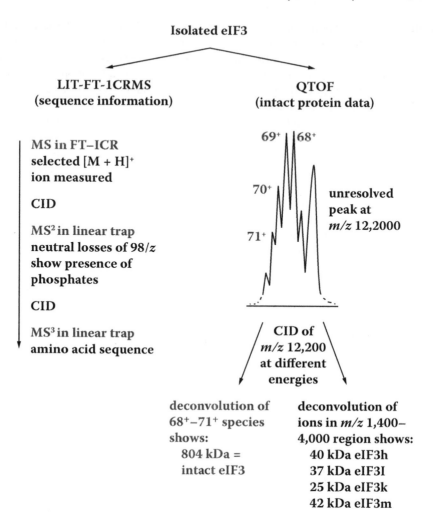

FIGURE 4.9   Analysis of eIF3. Component proteins (left) and intact noncovalent complex (right).

4.13   TISSUE IMAGING

Field: Tissue imaging.

Technique: MALDI.

Reference: P. Chaurand, D. S. Cornett, P. M. Angel, and R. M. Caprioli, From whole-body sections down to cellular level, multiscale imaging of phospholipids by MALDI mass spectrometry, *Mol. Cell. Proteomics* 10:1–11 (2011).

Background: Since its introduction ~15 years ago, imaging of tissues by MALDI has rapidly expanded to include peptides, proteins, multiple classes of lipids, as well as small molecules such as metabolites and xenobiotics. Tissues examined have ranged from frozen tissue sections to formalin-fixed paraffin-embedded sections, and have been obtained from a wide variety of sources, including from whole animals to specific tissues, e.g., kidney and eye lens, as well as normal and diseased states. While TOF (and TOF/TOF) remains the most frequently used technique, other approaches, including MS/MS using QTOF, trap-TOF, ion mobility, and FT-trap formats, have also been reported. This paper describes analysis of tissue sections from the whole body to subsections of tissues, down to the cellular level.

Methods: The MALDI process requires a matrix that is usually dissolved in a solvent prior to its addition to the sample. For tissue samples an even coating of the matrix can be obtained by spraying a solution onto the sample. A potential problem is that the solvent may affect the quality of the tissue image by solubilizing and causing migration of compounds within the sample. Solvent-free ways of obtaining an even coating of the matrix include sublimation and sieving it onto the tissue.

Sections, 10–12 µ thick, were cut from flash-frozen tissue, thaw-mounted on MALDI plates, and dried in a desiccator prior to deposition of the matrix. Sublimation of 2,5-dihydroxy-benzoic acid over ~4 min resulted in a homogenous 5–10 µ coating. Alternatively, the tissue was dry-coated with matrix that was passed through a 20 µ sieve for 20 min. Both methods were stable in the vacuum for 24 h, although back-sublimation is a risk that can create artificial analyte concentration gradients.

Results and discussion: Data for a section from a 1-day-old mouse pup with an area of ~224 mm^2 (200 µm resolution, 200 Hz laser) were acquired over a period of ~5.5 h. The spectra showed specific ions at high intensities in different tissues, e.g., m/z 770.5 in the eye, m/z 783.5 in the stomach, and m/z 810.6 in the adrenal gland. An example for a specific organ was the analysis of a mouse brain section using a resolution of 100 µm and both positive and negative ion acquisition; ~22 h was required for data acquisition. Again, certain ions could be assigned to specific regions of the tissue sections, e.g., m/z 734.6 in the cerebral cortex and striatum, m/z 741.6 in the ventricles, and m/z 826.7 in the striatum and corpus callosum (Figure 4.10).

Information at close to the cellular level provided data with very high spatial resolution, but there were limits caused by the size of the laser spot and control of the sample rastering. Data from a prototype instrument, with a laser focused to <5 µ diameter, yielded resolution down to 4.8 µ in brain tissue, enabling the localization of m/z 825 and m/z 851 ions within the nerve bundles of the striatum. Examination of the aortic valve from the heart of an adult mouse provided nonoverlapping spots with outer diameters of 6.5 µ, indicating that data could be obtained at a cellular level (collected for 6.5 h over a 540 × 1,100 µ area). The heart valve consisted of an outer layer of endothelial cells and an internal single cell layer of vascular interstitial cells. The cusp of the valve was three cells wide and showed ions at m/z 741 and m/z 846 but not at m/z 600; all three ions appeared in the hinge region of the valve (Figure 4.10). These results illustrate the potential for investigating the functioning of normal tissue as well as the changes that occur in abnormal (diseased) tissues, all at the cellular level.

The advances in imaging whole bodies and cellular structure illustrate the progress made but many aspects still require investigation. For instance, can high resolution MS techniques (FT-based systems) enable the separation and mapping of isobaric masses in tissues? A complementary method to high-resolution MS for deconvoluting isobaric species would be to use MS/MS to track fragments characteristic of specific classes of compounds. While this can be problematic for species such as phospholipids, because of the need to know the particular transitions to monitor by MS/MS, tracking would be useful where transitions are predictable or known, e.g., in assessing the locations of drugs and their metabolites. There also remains the need for sophisticated bioinformatic tools to correlate disease states, histology, and mass spectrometric imaging.

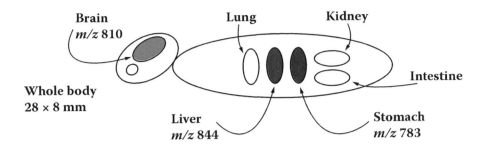

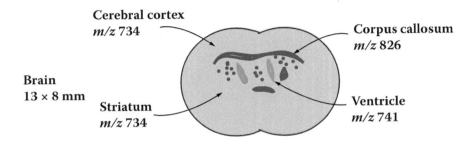

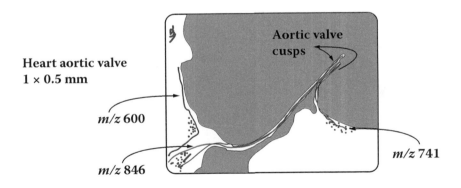

FIGURE 4.10 MALDI images of phospholipids in mouse tissues obtained at low (200 μ), medium (100 μ), and high (3 μ) resolution.

5 The Absolute Essentials

There are many aspects of mass spectrometry described/discussed throughout this book. Here is a condensation of what the authors consider the absolute essentials as a set of take-home messages. These are given in short phrases to refresh/remember, rather than in grammatically correct sentences.

Do not worry –
this is not a quiz.
Just a few bon mots
to be consumed each
morning with your breakfast.

5.1 GENERAL

Ion

- An atom, molecule, or fragment of a molecule that carries a positive or negative electrical charge

Important types of ions:

- Molecular, $[M]^{+\bullet}$; fragment (after breaking a covalent bond), $[X]^+$
- Protonated molecule (adding a proton to a molecule), $[M + H]^+$
- Adduct (adding an ionizing species, e.g., $[M + Na]^+$)
- Cation and anion, e.g., $[M + H]^+$, $[M - H]^-$
- Multiply charged, $[M + nH]^{n+}$

Isotopes:

- Species of elements with differing numbers of neutrons, e.g., ^{16}O, ^{17}O, ^{18}O; most isotopes are stable but some are radioactive, e.g., ^{14}C
- Some halogens, Cl and Br, and metals, e.g., Sn, Pt, and Pd, significantly alter the isotope patterns of small molecules

- Contributions of carbon isotopes may be significant for large biological compounds; there is a 1.1% chance that any given carbon in a compound will be a ^{13}C. Once there are more than 91 C atoms (proteins contain thousands of C), the peak(s) containing ^{13}C will be more intense than those containing only ^{12}C

m, *z*, AND *m/z*:

m, mass:

- Unit is dalton (Da); 1 Da = 1/12 of the mass of a ^{12}C atom

z, unit of charge:

- Unit of charge on a proton (positive) or an electron (negative), expressed (arbitrarily) in whole numbers, i.e., +1 or –1

m/z, mass-to-charge ratio:

- A measured property of an ion
- *m/z* is dimensionless; usually (but not necessarily) $z = 1$
- In electrospray, *z* increases with molecular mass

MASS

Nominal mass:

- Calculated, based on isotopic masses of the elements using integer masses, e.g., H = 1, O = 16

Exact mass:

- Calculated, based on isotopic masses using unified mass units based on C = 12.0000; e.g., H = 1.007825, O = 15.9949

Accurate mass:

- Measured mass of an analyte; value determined to the third or fourth decimal place

Monoisotopic mass:

- Calculated, based on the masses of the most frequently occurring isotopes

Molecular mass:

- Calculated, based on average atomic masses of the elements (all isotopes averaged), e.g., C = 12.0108 Da, O = 15.9994 Da

Isobaric mass:

- Empirical formulae that have the same nominal mass but different exact masses

MASS SPECTROMETER

MS or ms may mean (depending on context):
- Mass spectrometry, mass spectrometer, mass spectrometric, or mass spectrum (singular or plural)

Mass spectrometer:
- An analytical instrument that produces a beam of gas phase ions from samples, sorts the resulting mixture of ions according to their *mass-to-charge* (*m/z*) ratios using electrical or magnetic fields (or combinations thereof), and provides analog or digital output signals (*peaks*) from which the *m/z* and the *intensity* (abundance) of each detected ionic species may be determined

Mass spectrometers operate on the premises that:
- The sample is in the gas phase and is ionized (neutral molecules cannot be manipulated with electrical or magnetic fields)
- The ions are separated according to their *m/z* and then detected

Mass spectrometers obtain information on:
- *What* is present and *how much* is present by determining ionic masses and their intensities

Mass spectrometers measure:
- *m/z* not *m*

Components of a mass spectrum:
- A spectrum is a two-dimensional plot with *m/z* (not *m*) on the x-axis and intensity (usually relative) on the y-axis
- Depending on the ionization method, the molecular ion may be [M]$^+$, e.g., in EI, but adducts are formed in soft ionization, e.g., [M + H]$^+$, [M + adduct]$^+$ in ESI
- Other ions in a spectrum are fragments
- The most intense ion is the base peak
- Less intense ions (usually to the right of the major ions) are isotope peaks (mostly attributable to ^{13}C)

Diversity of MS:
- All spectroscopies, e.g., UV and IR, examine an invariate parameter of a molecule
- In mass spectrometry the ability to control where and how much energy is added to a molecule is widely variable and selectable, depending on the application, i.e., for structural characterization or quantification
- Added energy is determined/controlled in the ion source and in CID cells

Scope of MS

- Qualitative for confirmation/identification, quantitative for selected constituents

5.2 INSTRUMENT COMPONENTS

Sample introduction

Functions:
- Produce vapors from samples; introduce a sufficient quantity into the ion source so that its composition represents that of the original sample

Polarity is important for analytes of ≤1 kDa:
- Affects the selection of the sample introduction method
- Depends on the number of primary amino, hydroxyl, and acidic groups

Chromatographic separation of mixtures: GC, LC, and CE

Gas chromatography (GC):
- Samples must be nonpolar and remain stable when volatilized in helium (up to 320°C)
- Good to ~600 Da

Liquid chromatography (LC):
- Polar samples acceptable; atmospheric pressure ionization (API) methods enable interfacing with reversed phase columns
- Components are separated as the mobile phase is changed from aqueous to organic, with polar compounds eluting first
- Good to 100 kDa

Capillary electrophoresis (CE):
- Specific form of LC
- Efficient for large biomolecules
- Low flow rates make interfacing difficult

Semipermeable polymer membrane:
- Different permeabilities enable passage of volatile organic compounds through a silicone membrane (relative to water or inorganic gases)
- Selective introduction of certain analytes present in aqueous or gaseous streams

Other sample introduction methods:
- Batch inlet
- Direct insertion probe
- Atmospheric solids analysis probe

ION SOURCES

Functions:
- Produce an ion beam representative of the sample; then form, shape, and eject the beam into the mass analyzer

Hard ionization, such as electron ionization (EI):
- Imparts more energy than needed to form molecular ions
- Extensive fragmentation occurs because the excess energy causes dissociation of bonds; molecular ion may be lost

Electron ionization (EI):
- Takes place inside vacuum chamber
- Energetic (70 eV) electron beam from a heated filament impacts vaporized analytes, displaces an electron, and generates radical cations that frequently fragment
- Reproducibility of spectra is excellent; detection limits are low
- Mass limit 1 kDa
- Mostly used in GC-MS

Soft ionization (CI, API, MALDI):
- Added energy is sufficient only to generate adducted ions, e.g., $[M + H]^+$
- Ions formed are protonated molecules (not protonated molecular ions)

Energy transfer:
- Highest to lowest: EI > CI ≈ APCI ≈ APPI > MALDI ≈ ESI

Currently most important:
- ESI and MALDI, especially for biological (polar) analytes

Chemical ionization (CI):
- Takes place inside vacuum chamber
- Electrons from heated filament react with a reagent gas (e.g., methane) forming ions that in turn transfer a proton to the analytes, producing protonated molecules, $[M+H]^+$
- Particularly useful for quantification
- Mass limit 1 kDa
- Mostly used in GC-MS

Atmospheric pressure ionization (API):
- Includes ESI, APCI, and APPI
- Ionization takes place outside vacuum chamber
- Enables simultaneous vaporization and ionization of polar compounds
- Predominant ions are $[M + H]^+$ and $[M + adduct]^+$

Electrospray ionization (ESI):

- Analytes in solution are sprayed from a stainless tube held at 2–5 kV, droplets are dried, ions are released through charge repulsion
- Favors most polar analyte in a mixture
- Molecules of >1 kDa form multiply charged ions appearing as an envelope (skewed to lower m/z) that must be deconvoluted (transformed) to obtain the molecular mass of the neutral, zero charge state

Atmospheric pressure chemical ionization (APCI):

- Equivalent of CI at atmospheric pressure
- Corona discharge is used for ionization (instead of electrons from a filament)
- More uniform than ESI, but analytes must be polar
- Mass limit 1 kDa

Atmospheric pressure photo-ionization (APPI):

- Similar to APCI
- UV source used instead of corona discharge
- Extends ionization to compounds that are less polar than can be ionized with APCI, e.g., aromatic hydrocarbons

Surface ionization methods:

- Take place outside vacuum chamber at ambient pressure and temperature
- Include DESI (and related methods) and DART
- Mainly qualitative analysis of compounds adsorbed on surfaces

Desorption electrospray ionization (DESI):

- Spray of charged droplets (methanol, water) bombards a surface, ejecting and ionizing analytes
- No sample preparation, analysis *in situ*
- Cannot be combined with chromatography

Direct analysis in real time (DART):

- Ionization using an excited gas stream (He) that bombards the sample; simple, rapid
- Cannot be combined with chromatography

Matrix-assisted laser desorption/ionization (MALDI):

- Sample is mixed with ~1,000-fold excess of a special matrix (aromatic acid)
- Spotted and dried on a metal plate, placed in vacuum chamber
- Matrix absorbs the energy when irradiated with a UV laser; analyte and matrix evaporate
- Ions (singly charged) are formed at the surface or during evaporation
- Almost always analyzed using TOF

MASS ANALYZERS

Function:

- Separate ions according to their m/z

Classification of analyzers:

- Based on whether or not the operating parameters must be altered to observe ions with differing m/z values
- In scanning instruments (quadrupoles, traps, magnetic) electric or magnetic fields must be varied continuously to obtain spectra
- In nonscanning analyzers (TOF, FT-based) spectra are collected without changing operating parameters

Quadrupole (Q):

- Molybdenum rods (10–20 cm long, 1 cm diameter) placed in a square with opposite pairs connected electrically
- Mass separation (filtering) takes place when ions oscillate in the field produced by superimposed rf and dc voltages
- For any given field only one specific m/z has a stable trajectory and can move to the detector; all other ions describe unstable paths and discharge onto the rods
- Mass range is scanned by changing rf and dc voltages progressively while keeping their ratio constant
- Selected ion monitoring (SIM) uses a single rf/dc field
- Qs have unit resolution, mass range is to 4 kDa
- Compatible with GC-MS and LC-MS
- Quadrupole (q): in rf only mode acts as ion transmission device

Time-of-flight analyzer (TOF):

- Pulses of ions are accelerated from the source into an analyzer tube where the time is measured for an ion to travel through a field free region to the detector
- The time-of-flight is a function of the momenta of ions, and therefore of their m/z; the acceleration voltage, and thus the kinetic energy (momentum), is the same for all ions, so those with the lowest m/z will travel fastest and arrive at the detector first, followed by the sequential arrival of ions with successively higher m/z
- Delayed extraction reduces spatial distribution by aligning and concentrating ions at an electronic gate prior to their release into the analyzer (MALDI)
- Orthogonal injection is used to create ion pulses in ESI
- Reflectrons improve resolution and mass measurement accuracy by accounting for differences in the energy of ions that have the same empirical formula
- Resolution to 60,000
- Upper mass limit 350 kDa, theoretically unlimited
- Compatible with GC-MS and LC-MS

Orbitrap:

- Electrostatic ion trap comprised of a spindle-shaped inner electrode and a split outer electrode; ions (from API) rotate around and oscillate along the inner electrode
- Oscillating ions come close to the split outer electrode, inducing a current, called a transient, that is interpreted using Fourier transform (FT) analysis
- No superconducting magnet
- Resolution up to 250,000, accurate mass measurement, dynamic range $>10^3$
- Compatible with LC-MS but not at very high resolution

Fourier transform ion cyclotron resonance (FT-ICR):

- Based on the fact that ions move in circular paths in a magnetic field
- Frequency of rotation depends on the m/z of ions and strength of the magnetic field; direction of ion motion is such that a positive ion traveling through a magnetic field is set at 90° to the trajectory curves in a clockwise direction; radius of the curve is proportional to ionic mass (left-hand rule)
- Resolution up to 3 million, very accurate mass measurement (<1 mDa)
- Not compatible with LC-MS at highest resolution

Ion mobility separator (IMS):

- While MS characterizes ions according to m/z, IMS adds another dimension by enabling investigation of the shape of ions
- Ions enter a cell carried by voltage waves but counteracted by a flow of nitrogen
- Of two isomers, the one with more open form (greater cross-sectional area) will be slowed by the counterflow and emerge later than its more compact counterpart

Magnetic sector:

- Ions are accelerated by a fixed voltage; a specified angle of movement is required for ions to reach the detector
- Scanning the strength of the magnetic field systematically allows each ion with a given m/z to achieve the necessary angle of deflection and reach the detector
- Scan times are slow
- Used only for specific applications

Multianalyzer system (MS/MS):

- Now common to combine two or more analyzers within a single instrument (MS/MS) to improve and extend analytical capabilities
- Tandem-in-space combinations involve similar analyzers, tandem instruments, or use mixed types, hybrid instruments
- Common tandem MS/MS: triple quadrupole (QqQ) and TOF-TOF
- Common hybrid MS/MS: QTOF, LIT-orbitrap, and LIT-FTICR
- Tandem-in-time analyzers use the same analyzer twice (QIT and LIT)

- MS/MS investigates both mass and structure of ions: first analyzer is used to select an ion of interest that is passed, usually after fragmentation, into the second analyzer, often of higher resolving power, providing accurate mass measurement

Triple quadruple (QqQ): most versatile of MS/MS instruments:
- Used for product ion, precursor ion, and neutral loss scans as well as for selected reaction monitoring

Quadrupole ion trap (QIT):
- Three-dimensional Q; all ions maintained in stable trajectory (complex sinusoidal path) within the trap
- Selected *m/z* values are detected as they are rendered unstable by an applied voltage and ejected onto the detector
- Specified ions maintained in the trap with stable trajectories for sequential MS/MS analyses (MS^n)
- Limited number of ions because of space-charge effects

Linear ion trap (LIT):
- Similar to QIT, but ions oscillate in linear fashion along the length of a quadrupole
- Linear format removes most crossover points where major space-charge problems are encountered in QIT, thus improving sensitivity because more ions stored

Quadrupole time-of flight (QTOF):
- Quadrupole, used in Q or q mode, transmits ions to CID cell; resulting fragments are analyzed by TOF
- Accurate mass data

Linear ion trap–orbitrap:
- LIT can be used for MS or MS/MS
- Ions from LIT pass to orbitrap for accurate mass measurement
- May have second collision cell for high-energy CID

Linear ion trap–FTICR:
- LIT can be used for MS or MS/MS
- Ions passed to ICR cell for accurate mass measurement or IRMPD/ECD (fragmentation methods in ICR cells)

ION DETECTORS

Function:
- Detection of ions emerging from the analyzer

Arriving ions form very small current:

- One ion per second is 1.6×10^{-19} A
- Ion currents usually 10^{-9} to 10^{-16} A
- In FT instruments ions induce a transient signal

Electron multiplier:

- Arriving ions instigate an electron cascade increasing the signal about a million-fold
- Resulting signal is amplified electronically and recorded by the data system

Types of multipliers:

- Continuous dynode, discrete dynode, multichannel plate (MCP); photomultiplier detector (Daly)

Orbitrap and ICR:

- Ion detection is based on generated image currents
- The passage of an ion close to a metal surface induces an alternating electric current (transient)
- Frequency is proportional to m/z of the ion
- FT analysis is used to deconvolute the transient and obtain spectra

VACUUM SYSTEM

Functions:

- Prevent loss of ions by collision with neutral molecules (air) in evacuated chambers
- Remove unreacted molecules from ion source to prevent memory effects, e.g., from chromatographic effluents

Mean free path:

- Average distance that an ion can travel before encountering a neutral molecule
- May be kilometers/miles in ICR, necessitating vacuum of 10^{-10} Torr

First stage of pumping (rough vacuum):

- Rotary or scroll pumps
- Initially removes atmospheric gases (to 10^{-2} Torr)
- Continuously removes exhaust gases from high-vacuum pumps

Second stage of pumping (high vacuum):

- Turbomolecular pumps (spinning at 50,000–90,000 rpm); 10^{-5} to 10^{-10} Torr

Differential pumping:

- Maintains necessary vacuum in different chambers of the instrument, e.g., between CI source and analyzer

DATA SYSTEMS

Functions:
- Control operational processes
- Acquire and process generated data (e.g., thresholding, centroiding)
- Interpret data locally
- Post-process data (databases, Internet)

5.3   PERFORMANCE PARAMETERS

RESOLUTION

Resolution:
- Ability to differentiate one mass from another
- Defined with respect to the mass and width of a peak at half its maximum height; e.g., if peak for an ion at m/z 800 is 0.05 wide resolution is 16,000

Resolving power:
- The inverse of resolution

Analyzers have different resolutions:
- Quadrupoles have unit resolution throughout mass range, e.g., m/z 200 from 201 and 2000 from 2001
- TOF, orbitrap, and ICRMS up to 60,000, 250,000, and 3,000,000, respectively

Resolution in TOF:
- Function of design
- Data collection time makes no difference
- Harder to achieve high resolution at low masses because peaks are narrow

Resolution in FT:
- Time dependent
- For example, in an ICR, to obtain a resolution of 500,000 at m/z 800 with a 9.8 Tesla magnet, a 5 s transient is required
- Not compatible with chromatography when peaks are a few seconds wide
- More difficult to reach high resolutions as the mass of the analytes increases

ACCURACY OF MASS MEASUREMENT

Accurate mass measurements:
- Easier at high resolution because isobaric (same nominal mass) species are resolved
- Resolution does not ensure accuracy, because the latter depends on the calibration of the mass scale

- FT-ICRMS provides highest accuracies using a combination of accurate calibration, continuous data collection, and very high resolution
- Achievable with analyzers having moderate resolutions, e.g., 5,000, if there are no interferences from adjacent peaks

Accurate mass (applications and limits):

- Obtain empirical formula to assist in identification of small molecules
- Assignment of amino acid sequences and deconvolution of overlapping chromatographic peaks in tryptic digests of proteins
- Number of possible empirical formulae increases dramatically with m/z
- Number of possible empirical formulae increases for a specified m/z when the number of elements is increased, e.g., adding S, Si, Cl, and Br to the C, H, O, and N

Difference between calculated (exact) and experimentally determined (accurate) masses:

- Expressed in millidaltons (mDa), millimass units (mmu), or in parts per million (ppm)

MAXIMUM MEASURABLE M/Z (UPPER MASS LIMIT)

- Highest observable for a given instrument
- Not directly related to resolution
- Poor resolution at high mass is undesirable

LIMITS OF DETECTION (LOD) AND QUANTIFICATION (LOQ)

LOD:

- Minimum amount of an analyte that can be observed reliably
- Often defined as 3× background noise

Instrument LOD:

- Minimum amount of analyte observable

Sample LOD:

- Minimum amount of analyte observable after extraction from a matrix

LOQ:

- Minimum amount of an analyte that can be quantified reliably
- Often defined as 3× LOD

5.4 TECHNIQUES AND STRATEGIES

SINGLE ANALYZER INSTRUMENTS

- Spectra for confirmation of identity or identification of unknowns
- Quantification (often with selected ion monitoring)

COLLISION-INDUCED DISSOCIATION (CID)

- Used in MS/MS

CID (fragment) ions:

- Generated in a collision cell between two analyzers; selected precursor ions from MS1 are accelerated into a collision (target) gas
- Multiple collisions additively increase the internal energy of the ions and cause fragmentation
- Product ions analyzed in MS2

Triple quadrupoles:

- Most important use of CID is in selected reaction monitoring (SRM) for quantification

Peptides:

- CID fragmentation of bonds yields b and y ions that enable sequencing
- Drawback: phosphate post-translation modifications (PTMs) are lost
- Alternative fragmentation methods, electron transfer dissociation (ETD), and electron capture dissociation (ECD) preserve phosphates
- Drawback: large multiply charged ($z > 2$) peptides required

Infrared multiphoton dissociation (IRMPD):

- Equivalent of CID in FT-ICRMS

QUANTIFICATION OF SMALL MOLECULES

Selected ion monitoring (SIM):

- Single quadrupoles
- Analyzer voltages set to monitor selected single m/z

Selected reaction monitoring (SRM):

- Triple quadrupoles
- Analyte m/z is selected in Q1 subjected to CID in q2 and a chosen m/z is monitored with Q3

Specificity and sensitivity:

- Better in SRM than in SIM

Calibration curves:

- Regression analysis used to fit equation to standards and determine unknown concentrations of analytes with known identities, e.g., pharmaceuticals

External standard method:

- Series of individual (blank) samples spiked with increasingly higher concentrations of analyte

Internal standard (IS) method:

- Similar to external method but adding the same amount of a structurally similar IS to each sample prior to extraction
- Compensates for recovery differences
- Stable isotope labeled analytes are the best IS

Standard addition method:

- Requires multiple sample aliquots of the sample
- First aliquot used as is
- Subsequent aliquots are supplemented with increasing (known) amounts of the analyte

COMMON CONTAMINANTS

Small molecule analysis:

- Phthalates (plasticizers), siloxanes (GC-MS), PEG

Biopolymer analysis:

- Trypsin peptides, keratin peptides (human or animal); siloxanes and PEG (from buffers)

IMAGING MASS SPECTROMETRY

- Frozen (10–20 µ) tissue sections analyzed by MALDI/TOF
- Location of analytes within tissue, currently reaching cellular level
- Many types of compound, most commonly lipids

PROTEOMICS: SEQUENCING

Bottom-up:

- After enzymatic digestion (trypsin), peptides are analyzed by LC-MS/MS
- Combination of partial sequences and peptide masses for two or three peptides sufficient for identification using databases (accessed using the Internet)
- Hundreds of proteins may be identified in single analysis

Top-down:

- Intact individual proteins analyzed by FT-ICRMS
- Amino acid residues are removed in sequence; data are complex, as all fragment ions form multiply charged envelopes
- Must account for ^{13}C
- Post-translational modifications are preserved

De novo:

- Used when there is no information available in databases
- Entire amino acid sequence must be derived from mass spectra
- Steps: digest multiple samples with several enzymes to produce different sets of peptides; determine m/z ratios; use MS/MS to obtain amino acid sequences of each peptide; cross reference data sets to obtain overall sequence of protein.

PROTEOMICS: QUANTIFICATION

- Relative rather than absolute quantification used most frequently

Stable isotope labeling by amino acids in cell culture (SILAC):

- *In vivo* technique
- Cells fed labeled amino acids

Isotope-coded affinity tag (ICAT):

- Isolated proteins labeled with cysteine-specific tag

Isobaric tag for relative and absolute quantification (iTRAQ):

- Peptides labeled after enzymatic digestion

Label-free:

- Uses internal standard (co-injected) in manner similar to quantification of small molecules

HYDROGEN-DEUTERIUM EXCHANGE

- Exposed protons replaced by deuterons
- Analysis carried out at 0 °C, pH 2.6
- Back exchange followed with altered physiological conditions; investigate rates of structural changes
- Determine which amino acids form the surfaces of proteins

OTHER BIOPOLYMERS

Nucleic acids:

- Sequenced similarly to proteins

Carbohydrates:

- Form complex tree-like structures
- Require MS^n for characterization

Wow! That was intense
— you were right that I needed
to read everything.
Does this mean I am no longer a novice?

6 Resources

Wow! This is really cool, an actual library. I had only heard about them and did not realize they still exist. Be careful who you tell about its location or the firemen from Fahrenheit 451 will be on their way.

6.1 BOOKS, JOURNALS, REVIEW ARTICLES, CLASSICAL PUBLICATIONS

An excellent source of new books about to be published is www.amazon.com (search: "books" and then "mass spectrometry"). This Web-site also includes a large number of books already released, usually with a description provided by the publisher and sometimes with thoughtful reviews by readers.

There has been a continuing *Series in Mass Spectrometry* published by Wiley-Interscience; some 20 books have already been published on a wide variety of subjects.

Below is a list of recommended books mostly published since 2002.

6.1.1 GENERAL

Mass Spectrometry: A Foundation Course. 2012.
  Downard, K. Publisher: Royal Society of Chemistry.
Mass Spectrometry Handbook. 2012.
  Lee, M. S. 2012. Publisher: Wiley.
Mass Spectrometry: Mass-to-Charge Ratio. 2012.
  Sahoo, U., Sen, A. K., and Seth, A. K. Publisher: Lambert Academic Publishing.

Mass Spectrometry: A Textbook. 2nd ed., 2011.
    Gross, J. Publisher: Springer Verlag.
Dictionary of Mass Spectrometry. 2010.
    Mallet, A. I., and Down, S. Publisher: Wiley.
Modern Mass Spectrometry. 2010.
    Schalley, C. A., and Springer, A. Publisher: Wiley.
Mass Spectrometry: Instrumentation, Interpretation, and Applications. 2008.
    Kraj, A., Desiderio, D. M., Nibbering, N. M., Ekman, R., Silberring, J.,
    Brinkmalm, A. M. Publisher: Wiley.
Fundamentals of Contemporary Mass Spectrometry. 2007.
    Dass, C. Publisher: Wiley.
Mass Spectrometry: Principles and Applications. 3rd ed., 2007.
    De Hoffmann, E., and Stroobant, V. Publisher: Wiley.
*Introduction to Mass Spectrometry: Instrumentation, Applications, and
    Strategies for Data Interpretation.* 4th ed., 2007.
    Watson, J. T., and Sparkman, O. D. Publisher: Wiley.
The Expanding Role of Mass Spectrometry in Biotechnology. 2nd ed., 2006.
    Siuzdak, G. Publisher: MCC Press.
Mass Spectrometry Desk Reference. 2nd ed., 2006.
    Sparkman, O. D. Publisher: Global View Publishing.
Understanding Mass Spectra: A Basic Approach. 2nd ed., 2004.
    Smith, R. M. Publisher: Wiley.
Mass Spectrometry Basics. 2003.
    Herbert, C. G., and Johnstone, R. A. W. Publisher: CRC Press.

6.1.2 FORMULAS AND DERIVATIONS

Mass Spectrometry: A Textbook, 2nd ed., 2011.
    Gross, J. H. Springer Verlag.
    An excellent source of theoretical background (e.g., quasi-equilibirum
        theory and energy considerations) as well as relatively simple math-
        ematical treatment of the important analyzers. References are given to
        original sources. Also, there are revealing problems and solution acces-
        sible in a special website. The text is at graduate level, requites some
        background in physics and mathematics.
*Practical Aspects of Trapped Ion Mass Spectrometry: Theory and
    Instrumentation*, Vol. IV, 2010.
    March, R. E. and Todd, F. J. Publisher: CRC Press.
    A very relevant book on quadrupoles and ion traps.
Time-of Flight Mass Spectrometry, 1997.
    Cotter, R. J. *Am. Chem. Soc.*
    An older book that describes the basic principles of TOF, and several inno-
        vative techniques, including orthogonal extraction, post source decay,
        and delayed extraction.

Two detailed review articles on Fourier transform mass spectrometry.

Marshall, A. G., Hendrickson, C. L. and Jackson, G. S. 1998. Fourier transform ion cyclotron resonance mass spectrometry: A primer. *Mass Spectrom. Rev.* 17:1–35.
Scigelova, M., Hornshaw, M., Giannakopulos, A. and Makarov, A. 2011. Fourier transform mass spectrometry. *Mol. Cell Proteomics*, 10.1074/mcp. M111.009431.

6.1.3 INSTRUMENTATION AND TECHNIQUES

Sample Preparation in Biological Mass Spectrometry. 2011.
  Ivanov, A. R., and Lazarev, A. V. Publisher: Springer.
Gas Chromatography and Mass Spectrometry: A Practical Guide. 2nd ed., 2011.
  Sparkman, O. D., Penton, Z., and Kitson, F. G. Publisher: Academic Press.
Electrospray and MALDI Mass Spectrometry: Fundamentals, Instrumentation, Practicalities, and Biological Applications. 2010.
  Cole, R. B. Publisher: Wiley.
Practical Aspects of Trapped Ion Mass Spectrometry: Theory and Instrumentation. Vol. IV, 2010.
  March, R. E., and Todd, J. D. Publisher: CRC Press.
Mass Spectrometry Imaging: Principles and Protocol. 2010.
  Rubakhin, S. S., and Sweedler, J. V. Publisher: Humana Press.
Ion Mobility Spectrometry—Mass Spectrometry: Theory and Applications. 2010.
  Wilkins, C. L., and Trimpin, S. Publisher: CRC Press.
Liquid Chromatography Time-of-Flight Mass Spectrometry. 2009.
  Ferrer, I., and Thurman, E. M. Publisher: Wiley.
Introduction to Modern Liquid Chromatography. 3rd ed., 2009.
  Kirkland, J. J., and Dolan, J. W. Publisher: Wiley.
Trace Quantitative Analysis by Mass Spectrometry. 2008.
  Boyd, R. K., Basic, C., and Bethem, R. A. Publisher: Wiley.
LC/MS: A Practical User's Guide. 2006.
  McMaster, M. C. Publisher: Wiley.
Liquid Chromatography–Mass Spectrometry. 3rd ed., 2006.
  Niessen, M. A. Publisher: CRC Press.
Quadrupole Ion Trap Mass Spectrometry. 2nd ed., 2005.
  March, R. E., and Todd, J. F. Publisher: Wiley.

6.1.4 APPLICATIONS

6.1.4.1 General

Mass Spectrometry in the Biological Sciences. 2012.
  Burlingame, A. L., and A. Carr, S. A. Publisher: Humana Press.
Mass Spectrometry in Structural Biology and Biophysics: Architecture, Dynamics, and Interaction of Biomolecules. 2012.
  Eyles, S. J., Desiderio, D. M., and Nibbering, N. C. M. Publisher: Wiley.

6.1.4.2 Protoeomics

Mass Spectrometry of Glycoproteins: Methods and Protocols. 2013.
    Kohler, J. J., and Patrie, S. M. Publisher: Humana Press.
Protein and Peptide Mass Spectrometry in Drug Discovery. 2011.
    Gross, M. L., Chen, G., and Pramanik, B. Publisher: Wiley.
Proteomics of Biological Systems: Protein Phosphorylation Using Mass Spectrometry Techniques. 2011.
    Ham, B. H. Publisher: Wiley.
Introducing Proteomics: From Concepts to Sample Separation, Mass Spectrometry and Data Analysis. 2011.
    Lovric, J. Publisher: Wiley.
Mass Spectrometry for Microbial Proteomics. 2010.
    Shah, N., and Gharbia, S. E. Publisher: Wiley.
Mass Spectrometry of Non-Covalent Complexes: Supramolecular Chemistry in the Gas Phase. 2009.
    Schalley, C. A., and Springer, A. Publisher: Wiley.
Computational Methods for Mass Spectrometry Proteomics. 2008.
    Eidhammer, I., Flikka, K., Martens, L., and Mikalsen, S. O. Publisher: Wiley.
Mass Spectrometry of Proteins and Peptides. 2nd ed., 2008.
    Lipton, M. S., and Pasa-Tolic, L. Publisher: Humana Press.

6.1.4.3 Other Omics

Mass Spectrometry of Nucleosides and Nucleic Acids. 2009.
    Banoub, J. H., and Limbach, P. A. Publisher: CRC Press.
Metabolome Analysis: An Introduction. 2007.
    Smdsgaard, J., and Hansen, M. E. Publisher: Wiley.

6.1.4.4 Pharmaceutical, Medical, and Clinical

Illicit Drugs in the Environment: Occurrence, Analysis and Fate Using Mass Spectrometry. 2011.
    Castiglioni, S., Zuccato, E., and Fanelli, R. Publisher: Wiley.
A Practical Guide to Implementing Clinical Mass Spectrometry Systems. 2011.
    Leaver, N. Publisher: ILM Publications.
Mass Spectrometry in Drug Metabolism and Disposition: Basic Principles and Applications. 2011.
    Lee, M. S., and Zhu, M. Publisher: Wiley.
Clinical Mass Spectrometry in Sports Drug Testing: Characterization of Prohibited Substances and Doping Control Analytical Assays. 2010.
    Thevis, M. Publisher: Wiley.
Application of Mass Spectrometry: Methods and Protocols. 2009.
    Garg, U., and Hammett-Stabler, C. A. Publisher: Humana Press.
Mass Spectrometry in Drug Metabolism and Pharmacokinetics. 2008.
    Ramanathan, R. Publisher: Wiley.
Mass Spectrometry in Medicinal Chemistry. 2007.
    Hofner, G., Mannhold, R., and Kubinyi, H. Publisher: Wiley-VCH.

Medical Applications of Mass Spectrometry. 2007.
  Vekey, K., Telekes, A., and Vertes, A. Publisher: Elsevier.
Mass Spectrometry in Cancer Research. 2002.
  Roboz, J. Publisher: CRC Press.

6.1.4.5 Miscellaneous

Mass Spectrometry in Food Safety. 2011.
  Zweigenbaum, J. Publisher: Humana Press.
Mass Spectrometry in Grape and Wine Chemistry. 2010.
  Flamini, R., and Traldi, P. Publisher: Wiley.
Organic Mass Spectrometry in Art and Archeology. 2009.
  Colombini, M. P., and Modugno, F. Publisher: Wiley.
Mass Spectrometry in Biophysics. 2005.
  Kaltashov, I. A., and Eyles, S. J. Publisher: Wiley.
Mass Spectrometry in Polymer Chemistry. 2002.
  Barner-Kowollik, C., Gruendling, T., and Weidner, S. Publisher: Wiley-VCH.

6.1.5 JOURNALS

Several journals dedicated to mass spectrometry are listed below:

European Journal of Mass spectrometry (EJMS): http://www.impublications.
  com/content/european-journal-mass-spectrometry
International Journal of Mass Spectrometry (IJMS): http://www.journals.
  elsevier.com/international-journal-of-mass-spectrometry/
Journal of the American Society for Mass Spectrometry (JASMS): http://www.
  springer.com/chemistry/analytical+chemistry/journal/13361
Journal of Mass Spectrometry (JMS): http://onlinelibrary.wiley.com/journal/
  10.1002/(ISSN)1096-9888c
Mass Spectrometry Reviews: http://onlinelibrary.wiley.com/journal/10.1002/
  (ISSN)1098-2787/issues
Rapid Communications in Mass Spectrometry (RCM): http://www.wiley.com/
  WileyCDA/WileyTitle/productCd-RCM.html

Papers that utilize mass spectrometry can also be found in many parts of the scientific literature. Important new methods or applications are often found in *Nature, Science, Analytical Chemistry*, and the *Proceedings of the National Academy of Sciences of the United States of America*.

6.1.6 SELECTED REVIEW ARTICLES

The journal *Mass Spectrometry Reviews* is probably the best single source of information for both the novice and experienced user. The reviews selected below are intended to illustrate the diversity of present-day mass spectrometry.

6.1.6.1 General and Sample Preparation

Duncan, M. W. 2012. Good mass spectrometry and its place in good science. *J. Mass Spectrom.* 47:795–809.

Valkenborg, D., Mertens, I., Lemière, F., Witters, E., and Burzykowski, T. 2012. The isotopic distribution conundrun. *Mass Spectrom. Rev.* 31:96–109.

Glish, G. L., and Vachet, R. W. 2003. The basics of mass spectrometry in the twenty-first century. *Nature Rev. Drug Disc.* 2:140–150.

6.1.6.2 Instrumentation

Lapthorn, C., Pullen, F., and Chowdhry, B. Z. 2013. Ion mobility spectrometry-mass spectrometry (IMS-MS) of small molecules: Separating and assigning structures to ions. *Mass Spectrom. Rev.* 32:43–71.

van Agthoven, M. A., Delsuc, M.-A., Bodenhausen, G., and Rolando, C. 2013. Towards analytically useful two-dimensional Fourier transform ion cyclotron resonance mass spectrometry. *Anal. Bioanal. Chem.* 405:51–61.

Donato, P., Cacciola, F., Tranchida, P. Q., Dugo, P., and Mondello, L. 2012. Mass spectrometric detection in comprehensive liquid chromatography: Basic concepts, instrumental aspects, applications and trends. *Mass Spectrom. Rev.* 31:523–559.

Hommerson, P., Khan, A. M., de Jong, G. G., and Somsen, G. W. 2011. Ionization techniques in capillary electrophoresis-mass spectrometry: Principles, design, and application. *Mass Spectrom. Rev.* 30:1096–1120.

Huang, M.-Z., Cheng, S.-C., Cho, Y.-T., and Shiea, J. 2011. Ambient ionization mass spectrometry: A tutorial. *Anal. Chim. Acta* 702:1–15.

Urban, P. L., Amantonico, A., and Zenobi, R. 2011. Lab-on-a-plate: Extending the functionality of MALDI-MS and LDI-MS targets. *Mass Spectrom. Rev.* 30:435–478.

Wilm, M. 2011. Principles of electrospray ionization. *Mol. Cell Proteomics* 10:1.

El-Aneed, A., Cohen, A., and Banoub, J. 2009. Mass spectrometry, review of the basics: Electrospray, MALDI, and commonly used mass analyzers. *Appl. Spectrosc. Rev.* 44:210–230.

Perry, R. H., Cooks, G. R., and Noll, R. J. 2008. Orbitrap mass spectrometry: Instrumentation, ion motion and applications. *Mass Spectrom. Rev.* 27:661–699.

6.1.6.3 Quantification

Rodriguez-Suarez, E., and Whetton, A. D. 2013. The application of quantification techniques in proteomics for biomedical research. *Mass Spectrom. Rev.* 32:1–26.

Bantscheff, M., Lemeer, S., Savitski, M. M., and Kuster, B. 2012. Quantitative mass spectrometry in proteomics: A critical review update from 2007 to the present. *Anal. Bioanal. Chem.* 404:939–965.

Lemeer, S., Hahne, H., Pachl, F., and Kuster, B. 2012. Software tools for MS-based quantitative proteomics: A brief overview. *Methods Mol. Biol.* 893:489–499.

6.1.6.4 MS/MS

Holman, S. W., Sims, P. F. G., and Eyers, C. E. 2012. The use of selected reaction monitoring in quantitative proteomics. *Bioanalysis* 4:1763–1786.

Picotti, P., and Aebersold, R. 2012. Selected reaction monitoring-based proteomics: workflows, potential, pitfalls and future directions. *Nat. Methods* 6:555–566.

6.1.6.5 Data Processing and Bioinformatics

Choi, H. 2012. Computational detection of protein complexes in AP-MS experiments. *Proteomics* 12:1663–1668.

Hartler, J., Tharakan, R., Köfeler, H. C., Graham, D. R., and Thallinger, G. G. 2012. Bioinformatics tools and challenges in structural analysis of lipidomics MS/MS data. *Briefings Bioinform.* 10:1093.

Nakayama, H., Takahashi, N., and Isobe, T. 2011. Informatics for mass spectrometry-based RNA analysis. *Mass Spectrom. Rev.* 30:1000–1012.

Schadt, E. E., Linderman, M. D., Sorenson, J., Lee, L., and Nolan, P. 2010. Computational solutions to large-scale data management and analysis. *Nat. Rev. Genet.* 11:647–657.

6.1.6.6 Applications

Mechref, Y., Hu, Y., Garcia, A., and Hussein, A. 2012. Identifying cancer biomarkers by mass spectrometry-based glycomics. *Electrophoresis* 12:1755–1767.

O'Connell, T. M. 2012. Recent advances in metabolomics in oncology. *Bioanalysis* 4:431–451.

Senn, T., Hazen, S., and Tang, W. H. 2012. Translating metabolomics to cardiovascular biomarkers. *Prog. Cardiovasc. Dis.* 55:70–76.

Thevis, M., and Volmer, D. A. 2012. Recent instrumental progress in mass spectrometry: Advancing resolution, accuracy, and speed of drug detection. *Drug Testing Anal.* 4:242–245.

Yao, Z.-P. 2012. Characterization of proteins by ambient mass spectrometry. *Mass Spectrom. Rev.* 31:437–447.

Castellino, S., Groseclose, M. R., and Wagner, D. 2011. MALDI imaging mass spectrometry: Bridging biology and chemistry in drug development. *Bioanalysis* 3:2427–2441.

Gilliespie, T. A., and Winger, B. E. 2011. Mass spectrometry for small molecule pharmaceutical product development: A review. *Mass Spectrom. Rev.* 30:479–490.

Theodoridis, G., Gilka, H. G., and Wilson, I. D. 2011. Mass spectrometry-based holistic analytical approaches for metabolite profiling in systems biology studies. *Mass Spectrom. Rev.* 30:884–906.

6.1.7 CLASSIC PUBLICATIONS

There are three books in the literature on mass spectrometry that may be termed truly classic:

Rays of Positive Ray Electricity and Their Application to Chemical Analysis. 1913.
   Thomson, J. J. Publisher: Longman, Green.
Isotopes. 1923.
   Aston, F. W. Publisher: Arnolds.
Mass Spectra and Isotopes. 2nd ed., 1940.
   Aston, F. W. Publisher: Arnold.

Aston's two books should indeed be read by those wishing to choose mass spectrometry as their profession. His somewhat personalized description of instruments, experiments, and evaluations of results leaves one with the impression that Aston knew everything that could be known about mass spectrometry, that is, at that time; today he might well be considered a (promising) novice.

The American Society for Mass Spectrometry has been republishing "old" books in a series entitled *Classic Works in Mass Spectrometry.* A few selected subjects include:

Electron Impact Phenomena.
   Fields, F. H., and Franklin, J. L. Originally published in 1957 by Academic Press.
Mass Spectrometry and Its Applications to Organic Chemistry.
   Beynon, J. H. Originally published in 1960 by Elsevier.
Transport Properties of Ions in Gases.
   Mason, E. A. and McDaniel, E. W. Originally published in 1988 by Wiley.
Introduction to Mass Spectrometry: Instrumentation and Techniques.
   Roboz, J. Originally published in 1968 by Wiley.
Metastable Ions.
   Cooks, R. G., Beynon, J. H., Caprioli, G. R., and Lester, G. Originally published in 1973 by Elsevier.

For those interested in learning about mass spectrometry by tracing its history through original references, 10 selected classic papers are listed below. All are written in English and appeared in journals that are generally available in university libraries.

Thomson, J. 1911. Rays of positive electricity. *Phil. Mag.* 21:225–249.
Dempster, A. 1918. A new method of positive ray analysis. *J. Phys. Rev.* 11:316–324.
Aston, F. W. 1919. A positive ray spectrograph. *Phil. Mag.* 38:707–714.
Aston, F. W. 1920. The mass spectra of chemical elements. *Phil. Mag.* 39:611–625.
Dempster, A. J. 1922. Positive ray analysis of potassium, calcium, and zinc. *Phys. Rev.* 20:631–638.

Aston, F. W. 1925. Photographic plates for the detection of mass rays. *Cambridge Phil. Soc. Proc.* 22:548.

Aston, F. W. 1927. A new mass spectrograph and the whole number rule. *Proc. Royal Soc. A* 115:487–514.

Bartky, W., and Dempster, A. 1929. Paths of charged particles in electric and magnetic fields. *Phys. Rev.* 33:1019–1022.

Bainbridge, K. T. 1931. The blackening of photographic plates by positive ions of the alkali metals. *J. Franklin Inst.* 212:489–506.

Smythe, W. R., and Mattauch, J. 1932. A new mass spectrometer. *Phys. Rev.* 40:429–433.

6.2 MAJOR INSTRUMENT MANUFACTURERS

The major instrument manufacturers are listed below. There are many suppliers of components, e.g., specialized sources, multipliers, and pumps. Specialized software is also available from dedicated companies.

AB Sciex: http://www.absciex.com/
Agilent Technologies: http://www.home.agilent.com
Bruker (Daltonics): http://www.bruker.com/
Hitachi: http://www.hitachi-hitec.com/global/index.html
Jeol: http://www.jeol.com/; http://www.jeolusa.com/
Shimadzu: http://www.shimadzu.com/
Thermo Scientific: http://www.thermoscientific.com
Waters: http://www.waters.com/waters/home.htm

6.3 MASS SPECTROMETRY SOCIETIES, BLOGS, AND DISCUSSION GROUPS

American Society for Mass Spectrometry (ASMS), http://www.asms/org. The website of the ASMS is an excellent source of information about general and specialized meetings, publications, and a variety of functions provided by the society. Membership fees are reasonable and, indeed, it is a good idea for mass spectrometrists in the United States (and often in other countries) to maintain membership.

In addition to the ASMS, there is the European Society for Mass Spectrometry and at least 20 individual societies in various countries, e.g., United Kingdom, France, Germany, Canada, and India, each with its own website.

There are websites directed to mass spectrometrists, both beginners and advanced, discussing subjects ranging from technical problems to job offers in both industry and academia.

6.4   THE MASS SPECTROMETRIST AND THE INTERNET

Please note that website addresses change frequently and new websites often grow (and disappear) like mushrooms or apparitions in the wild. Accordingly, to prevent annoying frustration, we list below only a very limited number of actual sites (tested as of January 2013).

Indispensable resources on the Internet may be divided into four distinct sections: Wikipedia, tutorials, databases, and everything else.

6.4.1   WIKIPEDIA

To many novice mass spectrometrists (and to most advanced ones as well), Wikipedia (the free encyclopedia) is the first place to look for anything and everything. Wikipedia covers an astonishing number of mass spectrometry–related subjects, and the information is usually up-to-date. The articles are not really tutorials but are frequently informative and written in an easy-to-understand language. Yet, caution is suggested, as the quality of the information does vary due to the large number of contributors.

> Mass Spectrometry Wiki Mind Map: A blog on mass spectrometry, websites, discussion groups, mailing lists, and other links and items of interest to the mass spectrometry community.
> http://mass-spec.lsu.edu/blog/?p = 792: Posted by K. K. Murray, November 2012.

6.4.2   TUTORIALS

The already very large (and still rapidly increasing) number of tutorials on the Internet is a major resource for the novice (indeed, for all mass spectrometrists). All readers of this text are strongly encouraged to take advantage of this matchless (though often exasperating) resource. Attention is called to *videos*, particularly those that show (frequently with superb *animation*) the principles/operations of most components of mass spectrometers and most analytical techniques/strategies. Videos are nearly always in color, are usually only a few minutes long, and frequently include/ emphasize commercial aspects when prepared by instrument manufacturers. The other extreme of the tutorial landscape is the *lecture*. Often presented by first-class professors or scientists, short (few-minute) lectures deal with a selected aspect, usually a fundamental principle or special application, while long (up to an hour or more) lectures cover selected topics in detail.

Tutorials on the Internet range from outstanding to awful. A good bet (that does not always work) is to start with YouTube and select videos or tutorials made only after 2010. To whet the appetite, below is a list, in no particular order, of a few selected tutorials.

> There is a large website that includes information on an exceptionally wide variety of aspects of MS, both introductory and more advanced. Included are a series of lectures about most areas of MS, presented from different points of view by representatives of more than 20 major universities. Also

included are several databases, tools (e.g., to calculate isotope patterns), a variety of BioTools (e.g., post-translation modifications by peptide masses, including the respected Mowse program for protein identification from peptide mass data), a means to access articles in a number of journals, and even resources for job searching. However, clicking on an item may be frustrating, as availability keeps changing.

http://www.i-mass.com/guide/tutorial.html

Animation explaining some basic principles of MS:

http://www.dnatube.com/video/7235/

Dealing with basic principles of mass spectrometry, including magnetic analyzers:

http://www.youtube.com/watch?v=tOGM2gOHKPc

A very brief animation illustrating the principle of a quadrupole analyzer:

http://www.youtube.com/watch?v=8AQaFdI1Yow

Good YouTube videos on quadrupole analyzers:

http://www.youtube.com/watch?v=pjCun7QF19U

http://www.youtube.com/watch?v=L_k_AnlLn2s

Video explaining the principle of ion cyclotron resonance:

http://www.dnatube.com/video/7141/

Animation on the principle of desorption electrospray ionization (DESI) ion source:

http://www.prosolia.com/resources/videos/desi-mass-spectrometry

Principle of electrospray ionization—good but no animation:

http://www.waters.com/waters/nav.htm?cid=10073251&locale=en_US

What do you do if you have a sample from another planet and want to find out if it contains a certain molecule—one that will reveal if life can be sustained?

http://www.youtube.com/watch?v=_L4U6ImYSj0

Video lectures, part of a 28-lecture (1 h each) graduate course on all aspects of spectroscopy, e.g., Part 4 on basic theory of mass spectrometry and Part 5 on isotopes and high resolution.

http://www.youtube.com/watch?v=tvGkglvmf8s

http://www.youtube.com/watch?v=56wIpsZMYeU

Another recent video lecture (two parts) dealing with the use of mass spectrometry in proteomics, including identification, characterization, and quantification.

http://www.youtube.com/watch?v=xAyF6RIL1pM

An interesting simulator that permits one to observe mass spectra (of three compounds) on a console and modify them by changing parameters, e.g., resolution. There are both English and German versions.

http://vias.org/simulations/simusoft_msscope.html

6.4.3   DATABASES

There are numerous, absolutely essential databases, originally developed for proteins and nucleic acids, but by now covering many omics topics as well as more mundane items, such as thermochemical and spectroscopic properties of thousands of organic and small inorganic compounds. The three listed below direct one to numerous other related databases:

Swiss-Prot, important databases for proteomics: http://www.expasy.org/sprot
Database for nucleotide sequences: http://www.ncbi.nlm.nih.gov/Genbank
National Cancer Institute, including several tools and data sets: http://ncip.nci.
   nih.gov/

6.4.4 MISCELLANEOUS RESOURCES

Probably the most important among the large number of miscellaneous other
resources for the mass spectrometrist (novice or veteran) is PubMed (MEDLINE),
which deals with biomedical literature. This database is part of the search engine
of the National Center for Biotechnology Information. Here one can find complete
references, and often access to the full text, of just about all papers published on any
topics in mass spectrometry: http://www.ncbi.nlm.nih.gov/literature.

There are dozens of other websites describing/offering *software* (some free) for
a wide variety of *applications*, ranging from simple calculations of charge, molecular weight, and various properties of proteins to post-processing for both electrospray ionization (ESI) and matrix-assisted laser desorption/ionization (MALDI).
There are sites offering *protocols* (some tested, others wild looking), usually for
sample preparation.

The best way to conclude this section on resources, and this may sound corny, is
to wish the readers "good fishing." Surfing the Web is often impossibly frustrating,
but, in the end, success is virtually guaranteed.

Index

A

absolute essentials, 239

accuracy, 11, 16, 18, 22, 24–5, 83, 85, 94, 113, 115, 121, 122, 152, 154–6, 169, 179, 189, 212, 219, 245, 249, 261

accurate mass, 10, 30, 78, 117
    definition, 31, 117, 147, 233

accurate mass measurement, 25, 91, 96, 113, 117, 119, 146, 166, 196, 228, 240

acetaminophen, 2,12

adduct ion, 4, 59

alpha-cyano-4-hydroxy-cinnamic acid, 68, 195

amino acid, 23, 30, 163–5, 167–9, 170–1, 178, 188, 228, 230, 234, 250, 252, 253

ammonia, 48, 53–5, 64–5, 123, 188

analyzer
    ion cyclotron resonance, 18, 21, 71, 83–4, 86, 219, 246, 257, 260, 265
    magnetic, 1, 18, 25, 35, 71, 90–1, 97, 99, 103, 136, 199, 206, 245–6, 265
    orbitrap, 18–9, 21–2, 35–6, 71, 83–5, 96–7, 103, 113, 138, 141–2, 147, 173, 187, 198, 200–1, 246–9, 260
    quadrupole, 17–8, 20, 22–3, 30, 35–6, 71–3, 90–1, 94–5, 99, 104, 111, 113, 117, 134, 136, 145–6, 200, 209, 245
    time-of-flight, 1, 16, 17, 20, 22, 25, 36, 67, 71, 76, 79, 94, 95, 113, 136, 207, 245, 257

anion, 4, 46, 60, 172, 193, 239

anion attachment, 60

anthracenone, 118, 119, 122

APCI, 15, 31, 35–6, 39, 46, 47, 48, 62–4, 83, 112, 123, 130, 149, 159, 161, 200, 243–4

API, 18–9, 21–2, 35, 36, 71, 83–5, 96, 97, 103, 113, 138, 141–2, 147, 173, 187, 198, 200, 201, 246–9, 260

APPI, 15, 35, 36, 46–7, 63, 123, 200, 219, 243–4

Archimedian spirals, 105

Aston, F. W., 1, 262–3

atmospheric pressure chemical ionization (APCI), 11, 15, 19, 45, 48, 62, 207, 216, 244

atmospheric pressure ionization (API), 4, 11, 15, 39, 45, 206, 242, 243

atmospheric pressure photo-ionization (APPI), 15, 45, 63–5, 219, 244

atorvastatin, 213, 215

B

B (magnetic analyzer), 18, 33, 71, 91

b ions, 164–5, 173

background, 9, 28, 45, 59, 62–4, 144, 149, 156, 250
    ions, 22, 25, 108, 144

base peak, 7, 9, 47, 123, 130, 135, 241

batch inlet, 13, 36, 113, 136, 242

bioinformatic(s), 161, 196, 236, 261

biopolymer, 1, 17, 30, 108, 131–2, 162, 252

blank, 161, 196, 236, 261

books, 29, 255
    applications, 257
    classical, 262
    general, 255
    instrumentation, 257
    omics, 258
    pharmaceutical, medical and clinical, 258
    proteomics, 258

bottom-up, 166–9, 170, 252

brain, 235–7

blood plasma, 16, 31, 145, 179, 214

biomarker, 70, 146, 198, 222

bromine, 9, 121, 124–6

buying a mass spectrometer, 198

C

caffeine, 118–9, 122

calibration, 10, 13, 23, 36, 38, 90, 93, 108, 111–3, 121, 132, 135, 206, 249, 250–1

calibration curve, 150–6, 211, 214, 218

capillary columns, 12, 39, 40–2

capillary electrophoresis (CE), 12, 13, 39, 45, 61, 242, 260

carbohydrates, 162, 190, 193, 199, 253

carbon isotopes (^{13}C), 4, 7–9, 51, 125–6, 130, 209, 219, 220
    carbon count in small molecules, 126–7, 135, 154, 241
    affect on molecular mass of proteins, 10, 127, 169, 181, 182–3, 240, 252

carrier gas, 12, 40, 42

cation, 4, 14, 50, 59, 60, 70, 172, 231, 239

cation attachment, 59, 71

CE (capillary electrophoresis), 12–3, 39, 45, 61, 242, 260

centroid, 8, 73, 115, 228, 249